Constant Mean Curvature Surfaces in Homogeneous Manifolds

Flächen konstanter mittlerer Krümmung in homogenen Mannigfaltigkeiten
Zur Erlangung des Grades eines Doktors der Naturwissenschaften (Dr. rer. nat.)
genehmigte Dissertation von M.Sc. Julia Plehnert aus Bremen
Februar 2012 — Darmstadt — D 17

TECHNISCHE UNIVERSITÄT DARMSTADT

Fachbereich
Mathematik

Constant Mean Curvature Surfaces in Homogeneous Manifolds
Flächen konstanter mittlerer Krümmung in homogenen Mannigfaltigkeiten

Genehmigte Dissertation von M.Sc. Julia Plehnert aus Bremen

1. Gutachten: Prof. Dr. Karsten Groe-Brauckmann
2. Gutachten: Prof. Dr. Benoît Daniel

Tag der Einreichung: 3. Februar 2012
Tag der Prüfung: 18. April 2012

Darmstadt - D 17

Bibliografische Information der Deutschen Nationalbibliothek

Die Deutsche Nationalbibliothek verzeichnet diese Publikation in der Deutschen Nationalbibliografie; detaillierte bibliografische Daten sind im Internet über http://dnb.d-nb.de abrufbar.

ISBN 978-3-8325-3206-2

Logos Verlag Berlin GmbH
Comeniushof, Gubener Str. 47,
10243 Berlin
Tel.: +49 (0)30 42 85 10 90
Fax: +49 (0)30 42 85 10 92
INTERNET: http://www.logos-verlag.de

Contents

Zusammenfassung

Seit dem 18. Jahrhundert sind die Themen *Minimalflächen* und *H-Flächen* in der Mathematik aktuell und es gibt diverse Existenzresultate. Zu den wichtigsten gehört die Integraldarstellung von Minimalflächen im $\mathbb{R}^3$ von Weierstraß, die von Bryant und Bobenko auf Raumformen ausgeweitet wurde. Die verallgemeinerte Weierstraß-Darstellung für H-Flächen geht auf Dorfmeister, Pedit und Wu zurück. In den letzten Jahren lag das Interesse in H-Flächen in dreidimensionalen homogenen Räumen. Es gibt verschiedene Arbeiten, die sich mit der Darstellung von H-Flächen beschäftigen, siehe z.B. Abresch und Rosenberg ([AR05]). Diese Arbeiten sind sehr spezifisch; für den Existenzbeweis einer Fläche werden sehr viele Daten benötigt.

Oftmals besser zu handhaben ist die Lösung des Plateau-Problems und die Konjugiertenkonstruktion von Lawson. Diese Ansätze wurden von Karcher, Große-Brauckmann, Kusner und Sullivan mit neuen analytischen Methoden weiterentwickelt ([GKS07]). Viele H-Flächen in Raumformen, die systematisch studiert und klassifiziert worden sind, sind aus diesen Forschungen hervorgegangen. Jedoch benötigt dieser Weg Symmetrievoraussetzungen der Fläche und des umgebenden Raumes. Daniel hat in den letzten Jahren die Möglichkeit eröffnet, H-Flächen in homogenen Räumen auf diesem Weg zu untersuchen, ohne dass man sich auf Raumformen beschränken muss ([Dan07]).

In der vorliegenden Arbeit werden neue H-Flächen durch die Konjugiertenkonstruktion von Plateau-Lösungen konstruiert. Die von Daniel bewiesene erweiterte Lawson-Beziehung zwischen isometrischen Flächen in homogenen Räumen mit vierdimensionaler Isometriegruppe liefert keine vollständige Beschreibung erster Ordnung der konjugierten Fläche, sondern setzt lediglich ihre vertikalen Anteile (in Faserrichtung) in Relation. Dadurch wird die Konstruktion anspruchsvoller.

Die Arbeit ist in drei Teile untergliedert. Im ersten Teil werden die homogenen 3-Mannigfaltigkeiten näher untersucht, wobei drei verschiedene Ansätze verfolgt werden: Zuerst werden die homogenen 3-Mannigfaltigkeiten mit 4-dimensionaler Isometriegruppe als Riemannsche Faserungen mit geodätischen Fasern über zweidimensionalen Raumformen hergeleitet. Der zweite Ansatz verläuft über metrische Lie-Gruppen. Mit einer Beschreibung der Modellgeometrien von Thurston endet der erste Teil.

Im zweiten Teil werden Definitionen und Eigenschaften von Hyperflächen in Riemannschen Mannigfaltigkeiten eingeführt. Anschließend folgt ein Kapitel über

Flächen konstanter mittlerer Krümmung in Riemannschen Faserungen mit geodätischen Fasern. Die Eindeutigkeit von Graphen mit konstanter mittlerer Krümmung wird dort bewiesen. In Kapitel 7 werden die geometrischen Eigenschaften von Schwesterflächen dargestellt, insbesondere wird dabei auf Schwesterkurven eingegangen. Der zweite Teil schließt mit Existenzergebnissen von Minimalflächen als Plateau-Lösungen und ihren Grenzwerten.

Die Hauptergebnisse der Dissertation werden im letzten Teil vorgestellt. Nach einer kurzen Diskussion bekannter Beispiele, die als Barrieren in den Konstruktionen verwendet werden, wird zunächst eine 2-Parameter-Familie von H-Flächen in $\Sigma(\kappa) \times \mathbb{R}$ für $H \in [0, 1/2]$ und $\kappa \leqslant 0$ konstruiert, die eine Drehsymmetrie und zwei verschiedene Enden hat. Die Beispiele, $H = 1/2$-Flächen in $\mathbb{H}^2 \times \mathbb{R}$, gehen aus Minimalflächen im Heisenberg-Raum hervor, die von vertikalen und horizontalen Geodätischen berandet werden. Besonders interessant aber auch anspruchsvoll ist die Konstruktion einer Fläche von Geschlecht 1. Neben der Diskussion der Geometrie müssen Periodenprobleme gelöst werden. Das heißt, bei der Wahl der Parameter, die den Fundamentalbereich der Fläche definieren, muss sichergestellt werden, dass sich die Fläche nach Schwarz-Spiegelung schließt. Für den Beweis wird die Eindeutigkeit der Fläche benötigt, die sich aus einer verallgemeinerten Grapheneigenschaft ergibt. Durch die Kontrolle der Krümmungen und Längen der Randkurven kann anschließend mit einem Zwischenwertargument die Existenz einer Fläche mit Geschlecht 1 beweisen werden.

Abstract

In the 18th century mathematicians started to study minimal and constant mean curvature surfaces (CMC surfaces) and since then there have been several existence results. One of the most important results is the integral representation of minimal surfaces in $\mathbb{R}^3$ by Weierstraß, which was extended by Bryant and Bobenko to certain space forms. The generalized Weierstraß representation for CMC surfaces is due to Dorfmeister, Pedit and Wu. In recent years the interest in CMC surfaces is growing specially in three-dimensional homogeneous spaces. There are several works dealing with their representation, see for example Abresch and Rosenberg ([AR05]). These works are very specific; evidence of the existence of a surface needs a sufficient amount of data.

Easier to handle are often the solution of the Plateau problem and the conjugate construction by Lawson. They were developed by Karcher, Große-Brauckmann, Kusner and Sullivan with new analytical methods ([GKS07]). Many CMC surfaces in space forms have emerged from this research and have been systematically studied and classified. However, this approach needs symmetry assumptions on the surface and the ambient manifold. In recent years Daniel has opened the possibility to study CMC surfaces in homogeneous manifolds as conjugate surfaces ([Dan07]).

In this work new CMC surfaces are obtained using the conjugate construction of Plateau solutions. The Daniel correspondence between isometric surfaces in homogeneous 3-manifolds with 4-dimensional isometry group does not give a complete first-order description of the conjugate. It simply relates their vertical components (in fibre direction).

One class of examples, MC-1/2-surfaces in $\mathbb{H}^2 \times \mathbb{R}$, is based on minimal surfaces in Heisenberg space, which are bounded by vertical and horizontal geodesics. Interesting but also demanding is the construction of a surface with genus 1. Besides discussing the geometry, period problems have to be solved. Namely, from the choice of parameters of the surface's fundamental piece, one has to ensure that the surface closes after Schwarz reflection about mirror planes. In the proof one has to show uniqueness, which results from a generalized graph property. By controlling curvatures and lengths of the boundary curves the existence of a genus 1 surface follows from an intermediate value argument.

This thesis is divided in three parts. In the first part we describe homogeneous 3-manifolds and analyse them in three different ways: First of all, we derive the homogeneous 3-manifolds with 4-dimensional isometry group as Riemannian fibrations with geodesic fibres over space forms. The second approach is due to metric Lie groups. We finish this part with the presentation of the Thurston geometries.

The second part starts with some general definitions and properties of hypersurface theory in Riemannian manifolds, followed by a chapter about CMC surfaces in Riemannian fibrations with geodesic fibres. Thus we prove uniqueness of CMC graphs. In Chapter 7 we discuss the geometric properties of sister surfaces and especially of sister curves. The existence of minimal surfaces as finite and infinite Plateau solutions concludes the middle part.

The main results of this work are given in the last part. But first we present some known examples of minimal and CMC surfaces in homogeneous manifolds, which we need as barriers in our construction. The construction of a $2k$-noid with dihedral symmetry in $\Sigma(\kappa) \times \mathbb{R}$, $\kappa \leqslant 0$ follows. We complete the thesis with the construction of a k-noid with genus 1 and MC $1/2$ in $\mathbb{H}^2 \times \mathbb{R}$, and the focus of our interest here is the solution of its two period problems.

It is a pleasure to thank everyone who made this thesis possible, including my advisor, Prof. Dr. Karsten Große-Brauckmann, who suggested this problem to me and always made himself available for stimulating discussions. I would also like to thank Prof. Dr. Benoît Daniel for being a valuable second reader and advisor. Moreover, I am indebted to Prof. Dr. Jan Hendrik Bruinier, Prof. Dr. Steffen Fröhlich, and Dr. Ivan Izmestiev for their role on the board of thesis examiners. Many thanks to Prof. Dr. Laurent Mazet, Prof. Dr. Harold Rosenberg, and Prof. Dr. Valério Ramos Batista, as well, for their support. I feel honored to have an artistic execution of my dissertation by Prof. Dr. Karl H. Hofmann. Thank you for the permission to use it on the title page.

In my daily professional life, I have the pleasure of working with a friendly group of colleagues. I appreciate the scientific exchanges with Tristan Alex and Miroslav Vrzina. Sybille Drexler has always a friendly smile and encouraged my work with her support. Nicole Lehmann was always there for me when I needed someone to talk to.

Thanks, Marko and Stefan for their work editing (parts of) this thesis and giving me helpful remarks.

Last but not least, I'd like to thank the Friedrich-Ebert-Stiftung for their financial and intellectual support over the past few years.

Part I. Homogeneous manifolds

After Bolyai and Lobatschewski proved the existence of the hyperbolic plane in the 19th century, the general theory of curved spaces was then developed by Gauß in dimension two and Riemann in arbitrary dimension. Their first physical application was given by Einstein. General relativity reveals gravitation as curvature of the space time, a four-dimensional Lorentzian manifold.

But the canonical manifolds in mathematics are space forms, i.e. simply connected manifolds with constant curvature. Up to scaling they are represented by $\mathbb{R}^n$, $\mathbb{S}^n$ and $\mathbb{H}^n$. A natural generalization of space forms are homogeneous manifolds. Their isometry group is still transitive as in a space form, but they are not isotropic anymore.

In the 80s Thurston conjectured that every oriented closed 3-manifold could be cut into pieces, each of which admits a homogeneous geometry. Thurston distinguishes all homogeneous 3-manifolds by their isometry groups, see Chapter 3. The conjecture was recently proven by Perelman. We are particularly interested in the homogeneous 3-manifolds with 4-dimensional isometry group. They are represented by the geometry of: The product spaces $\mathbb{H}^2 \times \mathbb{R}$ and $\mathbb{S}^2 \times \mathbb{R}$, the universal covering of $\mathrm{PSL}_2(\mathbb{R})$, the Berger spheres and the Heisenberg space Nil_3. The Berger spheres are not one of Thurston's model geometries, since their isometry group is not maximal.

There are different approaches to analyse homogeneous 3-manifolds with 4-dimensional isometry group. One can show that they are Riemannian fibrations over space forms. Conversely, Große-Brauckmann and Kusner showed that a Riemannian fibration over a space form with geodesic fibres results in a homogeneous manifold with 4-dimensional isometry group, see Chapter 1. Last but not least we present a very recent result by Meeks and Pérez ([MP11]). They proved that, except for the product manifolds $\Sigma(\kappa) \times \mathbb{R}$, $\kappa > 0$, every simply-connected, homogeneous Riemannian 3-manifold is isometric to a metric Lie group.

1 Riemannian fibrations with geodesic fibres

Let us consider a fibration $\pi\colon E \to \Sigma$ of a 3-dimensional complete Riemannian manifold $(E, \langle\cdot,\cdot\rangle)$ over an oriented surface Σ and a vector field ξ tangent to the fibre. We decompose the tangent space $\mathrm{T}_p E$ of the total space E into its horizontal and vertical parts. The *horizontal space* $(\mathrm{T}_p E)^h$ is defined by

$$(\mathrm{T}_p E)^h := \{v \in \mathrm{T}_p E : \langle v, \xi(p)\rangle = 0\},$$

the vertical space is given by $(\mathrm{T}_p E)^v := \ker(\mathrm{d}\,\pi_p)$.

We require the fibration to fulfil three properties: It has to be a *Riemannian* fibration with *geodesic fibres* over a 2-dimensional *space form*.

First of all, we consider a *Riemannian fibration*, i.e. $\mathrm{d}\,\pi_p\colon (\mathrm{T}_p E)^h \to \mathrm{T}_{\pi(p)}\Sigma$ is an isometry. This means that the horizontal spaces as well as the tangent space of the base manifold look the same. Furthermore, this implies that the horizontal lifts of geodesics in Σ are geodesics in E. For an arbitrary curve γ in Σ there is always a unique *horizontal lift* $\tilde{\gamma}$ if we fix $\tilde{\gamma}(0) = p \in E$, i.e. $\tilde{\gamma}'(t) \in (\mathrm{T}_{\tilde{\gamma}(t)} E)^h$, for all t.

Secondly, we require the fibres to be geodesics. By computing the Lie derivative, this leads to the following proposition.

Proposition 1.1 ([GK10]). *For a Riemannian fibration $\pi\colon E \to \Sigma$ with geodesic fibres, there exists an unit vector field ξ tangent to the fibres, which is a Killing field.*

We call the vector field ξ the *vertical vector field*. The translations along the fibres are isometries, hence the Killing field ξ generates a subgroup G of $\mathrm{Iso}(E)$. The integral curves of ξ define a principal bundle with connection 1-form $\omega(X) = \langle X, \xi\rangle$, see below the proof of Theorem 3.3 for details. The curvature form is $\Omega := \mathrm{D}\,\omega = \mathrm{d}\,\omega + \frac{1}{2}[\omega,\omega] = \mathrm{d}\,\omega$. The following equation holds for two arbitrary vector fields X, Y

$$\begin{aligned}
\Omega(X,Y) &= \mathrm{d}\,\omega(X,Y) = X\omega(Y) - Y\omega(X) - \omega([X,Y]) \\
&= \langle\nabla_X Y, \xi\rangle + \langle Y, \nabla_X\xi\rangle - \langle\nabla_Y X, \xi\rangle - \langle X, \nabla_Y\xi\rangle - (\langle\nabla_X Y, \xi\rangle - \langle\nabla_Y X, \xi\rangle) \\
&= \langle\nabla_X\xi, Y\rangle - \langle\nabla_Y\xi, X\rangle \\
&\overset{(*)}{=} 2\langle\nabla_X\xi, Y\rangle,
\end{aligned}$$

where $(*)$ follows from the fact that ξ is a Killing field. Moreover, the fact that ξ is Killing implies $\xi\Omega(X,Y) = 0$, i.e. Ω is constant along the fibers. Moreover, we claim: For $X^h = X - X^v$ we have $\Omega(X^h, Y^h) = \Omega(X,Y)$. Therefore Ω induces a 2-form $\underline{\Omega} = (\pi^{-1})^*\Omega$ on the base manifold Σ, this motivates Definition 1.2. To see the claim we compute

$$
\begin{aligned}
\Omega(X^h, Y^h) &= 2\langle \nabla_{X^h}\xi, Y^h\rangle \\
&= 2\langle \nabla_{(X-X^v)}\xi, Y - Y^v\rangle \\
&= 2(\langle \nabla_X\xi, Y\rangle \underbrace{-\langle \nabla_X\xi, Y^v\rangle}_{=\langle \nabla_{Y^v}\xi, X\rangle} - \langle \nabla_{X^v}\xi, Y\rangle + \langle \nabla_{X^v}\xi, Y^v\rangle) \\
&= 2\langle \nabla_X\xi, Y\rangle,
\end{aligned}
$$

because $\nabla_U\xi = 0$ for any vertical vector field U.

Since Σ is oriented, there exists a $\pi/2$-rotation J on $\mathrm{T}_y\Sigma$, it induces a $\pi/2$-rotation R on $(\mathrm{T}_p E)^h$.

Definition 1.2. Let $\pi\colon E \to \Sigma$ be a Riemannian fibration with geodesic fibres. Its *bundle curvature* τ is a map $\tau : \Sigma \to \mathbb{R}$ given by

$$
\tau(y) := -\frac{1}{2}\Omega(X, RX) = \frac{1}{2}\langle [X, RX], \xi\rangle,
$$

hence $[X,RX]^v = 2\tau(y)\xi$, where X is an arbitrary horizontal unit vector field along $\pi^{-1}(y)$.

We see that Ω measures the non-integrability of the horizontal distribution. The induced 2-form $\underline{\Omega}$ factorizes the natural volume form vol_Σ of Σ

$$
-\frac{1}{2}\underline{\Omega} = \tau\,\mathrm{vol}_\Sigma,
$$

since with an arbitrary unit vector x we have $\mathrm{vol}_\Sigma(x, Jx) = 1$ and $\underline{\Omega}(x, Jx) = \Omega(\tilde{x}, R\tilde{x})$.

The third requirement concerns the base manifold only. If we consider a two-dimensional space form $\Sigma^2(\kappa)$ as base, the Riemannian fibrations are classified as follows:

Theorem 1.3 ([GK10]). *Suppose a Riemannian fibration $\pi\colon E \to \Sigma(\kappa)$ has geodesic fibres, where E is a complete oriented simply connected Riemannian 3-manifold. Then E is a homogeneous manifold with a 4-dimensional isometry group and constant bundle curvature τ.*

We write $E(\kappa,\tau)$ for those spaces. The isometry group of $E(\kappa,\tau)$ depends on the signs of κ and τ, and is equivalent to the isometry group of one of the following Riemannian manifolds:

curv.	$\kappa<0$	$\kappa=0$	$\kappa>0$
$\tau=0$	$\mathbb{H}^2\times\mathbb{R}$	$\mathbb{R}^3$	$\mathbb{S}^2\times\mathbb{R}$
$\tau\neq 0$	$\widetilde{\mathrm{SL}}_2(\mathbb{R})$	$\mathrm{Nil}_3(\mathbb{R})$	(Berger-)$\mathbb{S}^3$

Another interpretation of the bundle curvature is the vertical distance of a horizontal lift of a closed curve:

Lemma 1.4 (Vertical distances). *Let γ be a closed Jordan curve in the base manifold $\Sigma(\kappa)$ of a Riemannian fibration with geodesic fibres $E(\kappa,\tau)$. With Δ defined by $\partial\Delta=\gamma$, we have*

$$d(\widetilde{\gamma}(0),\widetilde{\gamma}(l)) = 2\tau\,\mathrm{area}(\Delta),$$

where $\widetilde{\gamma}$ is the horizontal lift with $\pi(\widetilde{\gamma}(0))=\pi(\widetilde{\gamma}(l))$, area *is the oriented volume and $d(p,q)$ denotes the signed vertical distance, which is positive if $\overline{pq}$ is in the fibre-direction ξ.*

Proof. We consider an arclength parametrization $\gamma\colon[0,l]\to\Sigma(\kappa)$ and its horizontal lift $\widetilde{\gamma}$. Then $\widetilde{\gamma}(0)$ and $\widetilde{\gamma}(l)$ are contained in one fibre, i.e. $\pi(\widetilde{\gamma}(0))=\pi(\widetilde{\gamma}(l))$. Hence, there exists a vertical arclength parametrized curve c with $c(0)=\widetilde{\gamma}(l)$ and $c(v)=\widetilde{\gamma}(0)$. The union $c\cup\widetilde{\gamma}=:\widetilde{\Gamma}$ is a closed curve in $E(\kappa,\tau)$ and c' is parallel to ξ. If $c'=\pm\xi$, then

$$\pm v=\int_0^v\langle c',\xi\rangle=\int_0^v\langle c',\xi\rangle+\int_0^l\langle\widetilde{\gamma}',\xi\rangle,$$

since $\widetilde{\gamma}'$ is horizontal. By definition of the connection 1-form ω, the sum of the integrals is equal to $\int_{\widetilde{\Gamma}}\omega$. We apply Stokes's Theorem to get

$$\int_{\widetilde{\Gamma}}\omega=\int_{\widetilde{\Delta}}\mathrm{d}\,\omega=\int_{\widetilde{\Delta}}\Omega=\int_{\pi(\widetilde{\Delta})}(\pi^{-1})^*\Omega,$$

where $\widetilde{\Delta}$ is any lift of Δ, such that $\pi\colon\widetilde{\Delta}\to\Delta$ is one-to-one and $\partial\widetilde{\Delta}=\widetilde{\Gamma}$. Since $(\pi^{-1})^*\Omega$ is a 2-form on Σ, it factorizes the natural volume form vol_Σ

$$\int_{\pi(\widetilde{\Delta})}(\pi^{-1})^*\Omega=-2\int_{\Delta}\tau\,\mathrm{vol}_\Sigma=-2\tau\,\mathrm{area}(\Delta).$$

We conclude that

$$2\tau \operatorname{area}(\Delta) = \begin{cases} -v, & \text{if } c' = \xi, \\ v, & \text{if } c' = -\xi, \end{cases}$$

and $d(\tilde{\gamma}(0), \tilde{\gamma}(l)) = d(c(v), c(0)) = 2\tau \operatorname{area}(\Delta)$. □

2 Metric Lie groups

Another approach to homogeneous 3-manifolds is due to Milnor via metric Lie groups, see [Mil76]. Meeks and Peréz used this interpretation to classify homogeneous 3-manifolds. Herewith we give a short introduction to their work [MP11]. Let $(G,\cdot)$ denote a Lie group. For $a \in G$ consider the left-multiplication $l_a\colon G \to G, x \mapsto a \cdot x$.

Definition 2.1.
- A Riemannian metric on G is called *left-invariant* if for all $a \in G$ the map $l_a\colon G \to G$ is an isometry.
- A *metric Lie group* is a Lie group $(G,\cdot)$ together with a left-invariant metric $\langle\cdot,\cdot\rangle$.

With this definition Meeks and Peréz proved the following theorem:

Theorem 2.2. *Every simply-connected homogeneous 3-manifold is isometric to either a metric Lie group or to $\Sigma^2(\kappa) \times \mathbb{R}$ with $\kappa > 0$.*

To classify the homogeneous manifolds (besides $\Sigma^2(\kappa) \times \mathbb{R}, \kappa > 0$), we distinguish between *unimodular* and *non-unimodular* Lie groups. A Lie group G is unimodular if the left Haar measure is also right-invariant.

In general if G is a connected 3-dimensional metric Lie group, then we can choose an orientation for the Lie algebra $L(G)$ such that the cross product $\times\colon L(G) \times L(G) \to L(G)$ is defined. Moreover, it factorizes the Lie bracket $[\cdot,\cdot]\colon L(G) \times L(G) \to L(G)$. Milnor proved that the unique endomorphism $K\colon L(G) \to L(G)$ given by $K(X \times Y) = [X,Y]$ is self-adjoint if and only if the Lie group G is unimodular, see [Mil76, Lemma 4.1]. Therefore, for a positively oriented orthonormal K-eigenvector basis $\{E_1, E_2, E_3\}$ of $L(G)$, we have

$$[E_2, E_3] = c_1 E_1, \quad [E_3, E_1] = c_2 E_2, \quad [E_1, E_2] = c_3 E_3.$$

The constants $c_1, c_2, c_3 \in \mathbb{R}$ are called *structure constants*.

A change in the metric like bcE_1, acE_2, abE_3 is orthonormal (for $a, b, c \neq 0$) and transforms the structure constants as $(c_1, c_2, c_3) \to (a^2 c_1, b^2 c_2, c^2 c_3)$. Hence, the unimodular Lie group is determined up to isomorphism by the signs of (c_1, c_2, c_3). Moreover, a change of orientation changes (c_1, c_2, c_3) to $(-c_1, -c_2, -c_3)$. Milnor classifies the unimodular metric Lie groups by the signs of their structure constants, as follows:

constants	metric Lie groups
$+, +, +$	$SU(2)$
$+, +, -$	$\widetilde{\mathrm{SL}}_2(\mathbb{R})$
$+, +, 0$	univ. covering of rigid motions in $\mathbb{R}^2$
$+, -, 0$	Sol_3
$+, 0, 0$	$\mathrm{Nil}_3(\mathbb{R})$
$0, 0, 0$	$\mathbb{R}^3$

Meeks and Peréz showed that a non-unimodular metric Lie group G as in the theorem is isomorphic and isometric to a semi-direct product with the canonical metric:

Consider a group homomorphism

$$\sigma\colon \mathbb{R} \to \mathrm{Aut}(\mathbb{R}^2), \quad \sigma(t) = \sigma_t\colon \mathbb{R}^2 \to \mathbb{R}^2, \quad p \mapsto \sigma_t(p).$$

In the semi-direct product space $\mathbb{R}^2 \rtimes_\sigma \mathbb{R}$, the multiplication is defined by:

$$(p_1, t_1) * (p_2, t_2) := (p_1 \circ \sigma_{t_1}(p_2), t_1 + t_2),$$

where $\circ$ and $+$ are the operations on $\mathbb{R}^2$ and $\mathbb{R}$, respectively. For $A \in M_2(\mathbb{R})$, we look at the group homomorphism given by:

$$\sigma_t(p) = e^{tA}p, \quad (p_1, t_1) * (p_2, t_2) := (p_1 \circ e^{t_1 A} p_2, t_1 + t_2).$$

In general, for $A = \left(\begin{smallmatrix} a & b \\ c & d \end{smallmatrix}\right)$ we choose coordinates $(x, y) \in \mathbb{R}^2$, $z \in \mathbb{R}$ and define left-invariant vector fields on $\mathbb{R}^2 \rtimes_\sigma \mathbb{R}$ by taking derivatives with respect to (p_2, t_2) as follows:

$$E_1 = a_{11}(z)\partial_x + a_{21}(z)\partial_y, \tag{2.1}$$

$$E_2 = a_{12}(z)\partial_x + a_{22}(z)\partial_y, \tag{2.2}$$

$$E_3 = \partial_z, \tag{2.3}$$

where $(a_{ij}(z))_{ij} := e^{zA}$.

For any X, Y in the tangent space of the identity $\mathrm{T}_0(\mathbb{R}^2 \rtimes_A \mathbb{R})$, the *canonical left-invariant metric* $\langle\cdot,\cdot\rangle_A$ on $\mathbb{R}^2 \rtimes_\sigma \mathbb{R}$ is given by continuation of $\langle X, Y\rangle_{\mathbb{R}^3}$ by left-multiplication.

A Lie group is unimodular if and only if the endomorphism $\mathrm{ad}_X\colon L(G) \to L(G)$ given by $\mathrm{ad}_X(Y) = [X, Y]$ is traceless for all $X \in L(G)$. For a non-unimodular metric Lie group G one considers the *unimodular kernel* $K(G) = \{X \in L(G)\colon \mathrm{tr}(\mathrm{ad}_X) = 0\}$. If G is a metric Lie group isomorphic to $\mathbb{R}^2 \rtimes_A \mathbb{R}$, we may choose a basis $\{L_1, L_2\}$ of $K(G)$ such that, after scaling of G, we have $\mathrm{tr}(A) = 2$ and a unique $A = \left(\begin{smallmatrix} 1+a & -(1-a)b \\ (1+a)b & 1-a \end{smallmatrix}\right)$, $a, b \geqslant 0$. The cases $a = 0 = b$ and $a = 1$, $b \geqslant 0$ correspond to $\mathbb{H}^3$ and $\mathbb{H}^2 \times \mathbb{R}$, respectively. For details, see [MP11].

3 The eight Thurston geometries

This chapter describes the classification of 3-dimensional manifolds traced back to Thurston. It is based on [Sco83] and [Thu97]. Details may be found in [Alm03].

Definition 3.1.

- A *geometry* denotes a pair (M, G), where M is a differentiable manifold and G is a group, which acts transitively on M and has compact stabilizers G_p for each $p \in M$.
- Two geometries (M, G) and (M', G') are called *equivalent* if there exists a diffeomorphism $\varphi\colon M \to M'$ and an isomorphism $\psi\colon G \to G'$, such that

$$\varphi(g.p) = \psi(g).\varphi(p), \qquad \text{for all } g \in G \text{ and } p \in M.$$

- A geometry (M, G) is called *maximal* if G is not contained in any larger group that acts transitively on M and has compact stabilizers.
- We say that a geometry (M, G) *admits a compact quotient* if there exists a subgroup $H < G$ that acts on M as a covering group and has a compact quotient, i.e. $M \to M/H$ is a covering map and M/H is compact.

Remark 1. Consider a geometry (M, G). Let $\widetilde{M}$ be the universal covering of M. Then there exists a natural geometry $(\widetilde{M}, \widetilde{G})$, where $\widetilde{G}$ denotes the diffeomorphisms on $\widetilde{M}$ which are lifts of elements of G.

Therefore, we restrict our study to simply connected manifolds.

Lemma 3.2. *Let (M, G) be a geometry, then there exists a G-invariant Riemannian metric $\langle\cdot,\cdot\rangle$ on M.*

Proof. By the definition of a geometry (M, G), the stabilizers G_p are compact. Therefore, for $p \in M$, there exists a G_p-invariant inner product $\langle\cdot,\cdot\rangle_p$ on the tangent space $\mathrm{T}_p M$ by averaging any inner product $\langle\cdot,\cdot\rangle'$. Since G acts transitively, the inner product $\langle\cdot,\cdot\rangle_p$ can be extended to M and defines a G-invariant Riemannian metric $\langle\cdot,\cdot\rangle$ on M. □

We formulate the classification of 3-dimensional geometries as stated by Scott [Sco83].

Theorem 3.3. *Any maximal, simply connected, 3-dimensional geometry that admits a compact quotient is equivalent to one of the geometries* $(M, \mathrm{Iso}(M))$*, where* M *is one of the Riemannian manifolds* $\mathbb{R}^3$, $\mathbb{S}^3$, $\mathbb{H}^3$, $\mathbb{S}^2 \times \mathbb{R}$, $\mathbb{H}^2 \times \mathbb{R}$, $\widetilde{\mathrm{SL}_2}(\mathbb{R})$, Nil *or* Sol.

To structure the proof of the theorem we formulate some lemmata first.

Lemma 3.4. *Let* (M, G) *be a geometry with identity component of the isotropy group* $G_p^0 \cong \mathrm{SO}_2(\mathbb{R})$ *and* ξ *be a smooth unit vector field that is fixed under* G_p^0 *and* G_1*-invariant for a subgroup* $G_1 < G$ *with index* $(G\colon G_1) \leqslant 2$*. Then there exists a flow* ψ_t *on* M *generated by* ξ *that acts by isometries.*

Proof. Let P denote the plane distribution that is orthogonal to ξ, it is also invariant under G_1. We consider the flow ψ_t on M generated by ξ, i.e. $\psi : U \to M$ with

$$\frac{\mathrm{d}}{\mathrm{d}t}\psi(t,p) = \xi(p), \quad \psi(t,p) = p,$$

is defined on a suitable open subset $U \subset \mathbb{R} \times M$ with $\{0\} \times M \subset U$.

The flow preserves the distribution P, since P is invariant under G_1, and the group acts transitively on P: For $v \in P_p$ and $w = \mathrm{d}\psi_{t,p} v$ consider $g \in G_1$ with $\mathrm{d}g v = w$. We see directly

$$\begin{aligned}\langle w, \xi(\psi(t,p))\rangle &= \langle \mathrm{d}g v, \xi(\psi(t,p))\rangle \\ &= \langle v, \mathrm{d}g^{-1}(\xi(\psi(t,p)))\rangle \\ &= \langle v, \xi(p)\rangle.\end{aligned}$$

For $p \in M$ and $0 \neq w \in P_p$, there exists $A_t = A_t(p,w) > 0$ such that

$$\left\|\mathrm{d}\psi_{t,p} w\right\| = A_t \left\|w\right\|. \tag{3.1}$$

For each $v \in P_p$, we find $g \in G_p^0$ and $c > 0$ such that $c \cdot \mathrm{d}g_p v = w$. Therefore, we have

$$A_t \left\|w\right\| = A_t \left\|c \cdot \mathrm{d}g_p v\right\| = cA_t \left\|v\right\|, \quad \text{since } g \in G_p^0 \cong \mathrm{SO}_2(\mathbb{R}).$$

On the other hand we have

$$\left\|\mathrm{d}\psi_{t,p} w\right\| = \left\|\mathrm{d}\psi_{t,p} c \cdot \mathrm{d}g_p v\right\| = c\left\|\mathrm{d}g_p\, \mathrm{d}\psi_{t,p} v\right\| = c\left\|\mathrm{d}\psi_{t,p} v\right\|.$$

Together with equation (3.1) we get $\left\|\mathrm{d}\psi_{t,p} v\right\| = A_t \left\|v\right\|$ for all $v \in P_p$, i.e. $A_t = A_t(p)$.

It remains to show that $A_t = 1$ for all t. As ψ_t is the flow of the vector field ξ, $\mathrm{d}\psi_t$ preserves it. Hence, the volume form on M is multiplied by A_t^2. We will use the fact that M admits a compact quotient, i.e. there exists a subgroup $H < G$ such that M/H is compact and $M \to M/H$ is a covering map.

We claim that there exists $H_1 < G_1$ with these properties. Since $|G/G_1| \leqslant 2$, we have two possibilities. If $|G/G_1| = 1$, then $G_1 = G$ and so $H_1 = H < G = G_1$.

For $|G/G_1| = 2$ and $H \not< G_1$ we have $G = G_1 \cdot H$. Define $H_1 := H \cap G_1$. Since $G_1 \lhd G$ is normal, we conclude from the second isomorphism theorem that

$$G/G_1 \cong H/H_1 \quad \text{and} \quad |G/G_1| = |H/H_1| = 2.$$

The map $M \to M/H$ is a covering map with compact quotient and $M/H_1 \to M/H$ is a double covering map. Hence, M/H_1 is compact. Since M is simply connected we get a covering map $M \to M/H_1$ and $H_1 \subset G_1$.

The covering map induces the flow ψ_t on M/H_1. Since M/H_1 is compact, we know that $\mathrm{vol}(M/H_1) < \infty$, and the diffeomorphism ψ_t would multiply the volume with A_t^2. Therefore, we have $A_t = 1$. This showed that $\mathrm{d}\psi_t$ maps $\mathrm{T}_p M$ isometrically onto $\mathrm{T}_{\psi_t(p)} M$. Since p was arbitrary, the flow of ξ consists of isometries. The vector field ξ is then a Killing field. □

Lemma 3.5. *Let (M, G) be a geometry with $G_p^0 \cong \mathrm{SO}_2(\mathbb{R})$ and $\mathscr{F}$ a 1-dimensional foliation of M, which is G^0 invariant. Then $M/\mathscr{F}$ is a 2-dimensional manifold.*

Proof. It is sufficient to show that, for each $p \in M$, we can find a surface S_p such that S_p intersects each leaf in at most one point.

Let $\mathrm{T}_p M = \mathrm{T}_p^v M \oplus \mathrm{T}_p^h M$ be the decomposition of the tangent space into the vertical and horizontal parts induced by the foliation $\mathscr{F}$. For $p \in M$, let $V := V_p$ be a neighbourhood such that, for every $q \in V$, there exists a connected neighbourhood $U_q \subset \mathrm{T}_q M$, on which $\exp_q : U_q \to V$ is a diffeomorphism. Let $D \subset \left(P_p \cap U_p\right) \subset \mathrm{T}_p M$ be a disc of radius r at p such that $S_p := \exp_p(D)$ is transversal to $\mathscr{F}$. We claim that S_p satisfies the desired property.

We prove by contradiction: For $q \in S_p$ we consider its leaf F_q and assume that there exists $s \neq p$ with $S_p \cap F_q = \{s\}$. We look at $v := \exp_q^{-1}(s) \in U_q$ and have $v \in \mathrm{T}_q M \setminus \mathrm{T}_q^v M$ because for $v \in \mathrm{T}_q^v M$ we get $\exp_q(v) \in L_q \cap S_p$. Since exp is a diffeomorphism, we have $s = p$. Therefore, any $g \in G_q^0 \setminus \{\mathrm{id}\}$ implies $v \neq \mathrm{d}g_q v \in U_q$. We get a contradiction by

$$g.s = g.\exp_q(v) \overset{(*)}{=} \exp_q(\mathrm{d}g_q v) \neq \exp_q(v) = s$$

and the fact that q and s are on the same leaf with the same stabilizer $G_q^0 = G_s^0$, i.e. for $g \in G_q^0$ we have $g.s = s$. To see $(*)$, one may consider $\frac{\mathrm{d}}{\mathrm{d}t} g.\exp_q(tv)\Big|_{t=0} = \frac{\mathrm{d}}{\mathrm{d}t}\exp_q(\mathrm{d}g_q tv)\Big|_{t=0}$. Hence, $M/\mathscr{F}$ is a 2-dimensional manifold. □

Proof of Theorem 3.3. Let (M,G) be a geometry as required. By Lemma 3.2 we can consider a G-invariant Riemannian metric $\langle\cdot,\cdot\rangle_M$ on M. For $p \in M$, we look at the stabilizer G_p and denote its connected component of the unity by G_p^0. Since $\langle\cdot,\cdot\rangle_M$ is a G-invariant metric, G acts with isometries on $(M, \langle\cdot,\cdot\rangle_M)$. Each stabilizer G_p acts on $\mathrm{T}_p M$ and preserves the inner product. Therefore, G_p is a subgroup of $O_3(\mathbb{R})$. Since G_p^0 is connected, there are only three possibilities: It is either $\mathrm{SO}_3(\mathbb{R})$, $\mathrm{SO}_2(\mathbb{R})$ or trivial.

- $G_p^0 \cong \mathrm{SO}_3(\mathbb{R})$: Since the isometry group acts transitively, for p, $q \in M$ and for each pair of planes (σ_1, σ_2) with $\sigma_1 \subset \mathrm{T}_p M$ and $\sigma_2 \subset \mathrm{T}_q M$, there exists $\varphi \in \mathrm{Iso}(M)$ with $\varphi(p) = q$ such that

  $$\mathrm{d}\varphi(\sigma_1) = \sigma_2.$$

  Therefore, M has constant sectional curvature K. Up to scaling $\langle\cdot,\cdot\rangle_t = t\langle\cdot,\cdot\rangle_M$, for $t > 0$, the manifold is $\mathbb{R}^3$, $\mathbb{S}^3$ or $\mathbb{H}^3$ for $K = 0$, > 0 or < 0, respectively; see [GHL04, p. 160] for details.

- $G_p^0 \cong \mathrm{SO}_2(\mathbb{R})$: Let $L_p \subset \mathrm{T}_p M$ be the line which is fixed under the actions of G_p^0 and $P_p \subset \mathrm{T}_p M$ its complement. Since G_p^0 is a subgroup of G_p, both L_p and P_p are also fixed under G_p.

  The map $L\colon M \to \mathrm{T}M$, $p \mapsto L_p$, generates a line field on M. Since M is simply connected, there is an orientation that admits a smooth unit vector field ξ with $\xi(p) \subset L_p$. The vector field does not need to be G-invariant, but there is a subgroup $G_1 < G$ under which it is invariant and its index satisfies $(G\colon G_1) \leqslant 2$. Since ξ and P_p are both G_1-invariant, they preserve every manifold covered by M, whose covering group is contained in G_1. We call ξ the vertical vector field and P the horizontal distribution.

  We consider the flow ψ_t on M generated by ξ, i.e. $\psi : M \times I \to M, I \subset \mathbb{R}$ with

  $$\frac{\mathrm{d}}{\mathrm{d}t}\psi_t(p) = \xi(p), \quad \psi_0(p) = p.$$

  By Lemma 3.4 ψ_t acts by isometries.

We consider the integral curves $c_p\colon I \to M$ of ξ, i.e. $c(0) = p$ and $\frac{\mathrm{d}}{\mathrm{d}t}c(t) = \xi(c(t))$. They form a 1-dimensional foliation $\mathscr{F}$ of M, which is G^0-invariant. It makes sense to consider the stabilizer G_F of a leaf $F \in \mathscr{F}$, because for $p \in F \subset \mathscr{F}$ and $g \in G^0_p$ we have

$$c_p(t) = \psi_t(p) = \psi_t(g.p) = g.\psi_t(p) = g.c_p(t) \in F,$$

because ψ_t commutes with g, as

$$\frac{\mathrm{d}}{\mathrm{d}t} g.\psi_t(p)) = \mathrm{d}g \frac{\mathrm{d}}{\mathrm{d}t}\psi_t(p) = \mathrm{d}g\xi(p) = \xi(g.p) = \frac{\mathrm{d}}{\mathrm{d}t}\psi_t(g.p).$$

ξ and g commute, since ξ is invariant under $g \in G^0_p$. For p, $q \in F$, we have $G_p = G_q$. Consider $Y := M/\mathscr{F}$. By Lemma 3.5 the quotient Y is a 2-dimensional manifold.

The flow ψ_t of ξ acts by isometries and we can decompose each $w \in \mathrm{T}M$ in two components, one is in ξ-direction and the other in the horizontal distribution P. Thus, we get a Riemannian metric $\langle\cdot,\cdot\rangle_Y$ on Y by

$$\langle v_1, v_2\rangle_Y^{[p]} := \langle w_1, w_2\rangle_M^p = \langle \mathrm{d}\psi_{t,p}w_1, \mathrm{d}\psi_{t,p}w_2\rangle_M^{\psi_t(p)},$$

where $v_i \in \mathrm{T}_{[p]}Y$ and $w_i \in \mathrm{T}_pM$. Since G acts transitively on M and preserves the distribution P, the metric on Y, induced by the natural projection $\pi\colon M \to Y$, defines an isometry between P_p and $\mathrm{T}_{\pi(p)}Y$ for every $p \in M$. Furthermore, G preserves the foliation $\mathscr{F}$ and all the leaves are diffeomorphic. So, they are either $\mathbb{R}$ or $\mathbb{S}^1$.

The group G descends to a group G_Y that acts transitively on Y. The manifold Y is simply connected, since M is its simply connected covering, which is also connected. Therefore, it must be a 2-dimensional space form $\Sigma^2(\kappa)$.

Claim: The foliation $\mathscr{F}$ defines an $\boldsymbol{F}$-principal bundle over Y, where $\boldsymbol{F}$ is $\mathbb{R}$ or $\mathbb{S}^1$.

The flow ψ_t defines an action of $\boldsymbol{F}$ on M. For every $p \in M$, we consider S_p as in the proof of Lemma 3.5 and its saturation $\mathrm{sat}(S_p)$, which is the union of all leaves that intersect S_p. With $U_p = \pi(S_p) \subset Y$, define

$$f : U_p \to S_p \subset \mathrm{sat}(S_p)$$

by $\pi \circ f(y) = y$, for all $y \in U_p$. The bundle charts

$$\varphi_{U_p} : \pi^{-1}(U_p) \to U_p \times \boldsymbol{F},$$

where F is the fibre $\mathbb{S}^1$ or $\mathbb{R}$, are defined by

$$\varphi(\psi_{\alpha t}(f(y))) = (\pi(f(y)), t),$$

where α is the length of the fibre $F_{f(y)}$ for $F = \mathbb{S}^1$ and $\alpha = 1$ for $F = \mathbb{R}$. It is F-equivariant with respect to the canonical action of F on $U_p \times F$. □

We have seen above that the line field L and the distribution P are G-invariant, i.e. $\forall p \in M$, $\forall g \in G$, $\forall w \in L_p$ and $\forall w \in P_p$ hold $\mathrm{d}g_p w \in L_{g.p}$ and $\mathrm{d}g_p w \in P_{g.p}$.

Claim: P defines a connection 1-form $\omega := \langle \xi, \cdot \rangle_M$.

If π_v denotes the vertical projection $\mathrm{T}_p M \to (\mathrm{T}_p M)^v$, we have $\omega(w) = \pi_v(w) = w^v$. We have to show that ω is G-equivariant. Namely,

$$(\forall p \in M)(\forall w \in \mathrm{T}_p M)(\forall g \in G) \quad \omega_p(\mathrm{d}g_p w) = \mathrm{d}g_p \omega_p(w) = \mathrm{d}g_p w^v.$$

Since P and L are G-invariant, we have

$$\mathrm{d}g_p w = \mathrm{d}g_p(w^v + w^h) = \mathrm{d}g_p w^v + \mathrm{d}g_p w^h \in \underbrace{(\mathrm{T}_p M)^v}_{\cong \mathbb{R} = \mathfrak{f}} \oplus (\mathrm{T}_p M)^h,$$

where $\mathfrak{f}$ denotes the Lie algebra of F. So we get

$$\omega_p(\mathrm{d}g w) = \mathrm{d}g \omega_p(w),$$

or with the pull-back notation $g^*\omega = \mathrm{d}g.\omega$. □

For this connection form we regard its curvature form $\Omega := \mathrm{D}\omega = \mathrm{d}\omega + \frac{1}{2}[\omega, \omega] = \mathrm{d}\omega$, where D denotes the total differential and d the exterior one. It is a 2-form on M with values in $\mathfrak{f} = \mathbb{R}$. The curvature form is G-equivariant as well, since

$$g^*\Omega = g^* \mathrm{d}\omega = \mathrm{d}(g^*\omega) = \mathrm{d}(\mathrm{d}g.\omega) = \mathrm{d}g.\mathrm{d}\omega = \mathrm{d}g.\Omega.$$

It means that G acts transitively on Ω and preserves ω, i.e. the plane distribution has constant curvature $\hat{\tau}$.

Now we look closer at the Riemannian metric $\langle \cdot, \cdot \rangle_M$ of the principal fibre bundle $M(\Sigma^2(\kappa), F, \pi)$. It was chosen to be G-invariant, i.e. $(M, \langle \cdot, \cdot \rangle_M)$ is homogeneous. As π induces an isometry between the horizontal space $\mathrm{T}^h M$ and $\mathrm{T}\Sigma^2(\kappa)$, it is a Riemannian fibration and the metric on the horizontal space is determined by the standard metric of $\Sigma^2(\kappa)$. Let R denote a $\pi/2$-rotation on the horizontal bundle induced by the orientation of $\Sigma^2(\kappa)$. Then the bundle curvature is given by

$$\hat{\tau} = \Omega(X,RX) = X.\underbrace{\langle \xi, RX\rangle_M}_{=0} - RX.\underbrace{\langle \xi, X\rangle_M}_{=0} - \langle \xi, [X,RX]\rangle_M = -\langle \xi, [X,RX]\rangle_M,$$

where $X \in \mathrm{T}^h M$ is unitary. Thus we have a 2-parameter family of metrics $\langle\cdot,\cdot\rangle_{(\kappa,\hat{\tau})} := \langle\cdot,\cdot\rangle_M$ depending on the base curvature κ and the bundle curvature $\hat{\tau}$ of M. We examine the different cases.

- $\hat{\tau} = 0$: If the curvature form is zero, ω is called a flat connection form and the horizontal distribution $\mathrm{T}^h M \subset \mathrm{T}M$ is involutive. Moreover, (M,ω) is isomorphic to $Y \times \boldsymbol{F}$ with the canonical connection, see [KN63, pp. 92]. Since M is simply connected, $\boldsymbol{F}$ has to be $\mathbb{R}$ and up to scaling of the metric we get three different manifolds $\mathbb{H}^2 \times \mathbb{R}$, $\mathbb{S}^2 \times \mathbb{R}$ and $\mathbb{R}^2 \times \mathbb{R}$ for $\kappa = -1$, 1 and 0, respectively. As the geometry is maximal, we end up with either $\mathbb{H}^2 \times \mathbb{R}$ or $\mathbb{S}^2 \times \mathbb{R}$.
- $\hat{\tau} \neq 0$: If the curvature form is not zero, the horizontal bundle $\mathrm{T}^h M \subset \mathrm{T}M$ is not integrable and M is a contact structure.
  * Let us first consider $\kappa < 0$. We have $\boldsymbol{F} = \mathbb{R}$, since otherwise the bundle $M(\Sigma^2(\kappa), \mathbb{S}^1, \pi)$ would be $\Sigma^2(\kappa) \times \mathbb{S}^1$, as $\Sigma^2(\kappa)$ is contractible and that would be a contradiction to the simply connectivity of M.

    A metric on $\Sigma^2(\kappa)$ defines a natural metric on the tangent bundle $\mathrm{T}\Sigma^2(\kappa)$ by the natural projection $p\colon \mathrm{T}\Sigma^2(\kappa) \to \Sigma^2(\kappa)$, as $\mathrm{d}p_w\colon \mathrm{T}_w(\mathrm{T}\Sigma^2(\kappa)) \to \mathrm{T}_z\Sigma^2(\kappa)$ is an isometry for $w \in \mathrm{T}_z\Sigma^2(\kappa)$ and $z \in \Sigma^2(\kappa)$. We may restrict the metric to the circle bundle $U_t\mathrm{T}\Sigma^2(\kappa)$, where t denotes the radius of the circles; it is a sphere-bundle over $\Sigma^2(\kappa)$. The group $\mathrm{PSL}_2(\mathbb{R})$ acts on $\Sigma^2(\kappa)$. One may check that this action is transitive and as the isotropy group of $\Sigma^2(\kappa)$ is $\mathrm{SO}_2(\mathbb{R})$. We know that $\Sigma^2(\kappa) \cong \mathrm{PSL}_2(\mathbb{R})/\mathrm{SO}_2(\mathbb{R})$. By considering the circle tangent bundle we get $U_t\mathrm{T}\Sigma^2(\kappa) \cong \mathrm{PSL}_2(\mathbb{R})$. Therefore, $\mathrm{PSL}_2(\mathbb{R})$ endowed with the metric of $U_t\mathrm{T}\Sigma^2(\kappa)$ is isometric to $\Sigma^2(\kappa)$.

    As $\mathrm{PSL}_2(\mathbb{R})$ is doubly covered by $\mathrm{SL}_2(\mathbb{R})$, then $\widetilde{\mathrm{SL}}_2(\mathbb{R})$ is its universal covering and also the one of $U_t\mathrm{T}\Sigma^2(\kappa) \cong \mathrm{PSL}_2(\mathbb{R})$. But the circle bundle is a $\mathbb{S}^1_t$-bundle over $\Sigma^2(\kappa)$; given that $\widetilde{\mathrm{SL}}_2(\mathbb{R})$ is simply connected it has to be a line bundle over $\Sigma^2(\kappa)$. The horizontal distribution in $U_t\,\mathrm{T}\Sigma^2(\kappa)$ induces a horizontal distribution in $\widetilde{\mathrm{SL}_2}(\mathbb{R})$. As $\tilde{p}\colon \widetilde{\mathrm{SL}_2}(\mathbb{R}) \to U_t\,\mathrm{T}\Sigma^2(\kappa)$, $(z,s) \mapsto (z, t\cos s, t\sin s)$ is a local isometry, this distribution is not integrable. Namely, the bundle curvature is not zero. Eventually we have $\widetilde{\mathrm{SL}_2}(\mathbb{R}) = M(\Sigma^2(\kappa), \mathbb{R}, \pi)$.

  - ⋆ For $Y = \mathbb{R}^2$ ($\kappa = 0$) we have a similar situation as above: our fibre $\boldsymbol{F}$ has to be $\mathbb{R}$. Therefore, M is a line bundle over $\mathbb{R}^2$ and is isometric to $\mathrm{Nil}_3(\mathbb{R})$.
  - ⋆ If $\kappa = 1$, then $Y = \mathbb{S}^2$. Its unit tangent bundle is $\mathrm{SO}_3(\mathbb{R})$. The corresponding universal cover is $\mathbb{S}^3$ and for G we would get those isometries of $\mathbb{S}^3$ preserving the Hopf fibration. But this is not a maximal geometry.

- $G_p^0 \cong \mathrm{id}$: It means that G^0 acts transitively and freely on M, and we may identify M with $G^0 = G^0/G_p^0$. Hence, M is a Lie group. We know that M admits a compact quotient, so we are looking for a 3-dimensional Lie group M and a subgroup $H < G$, which acts on M as covering group and has a compact quotient. Furthermore, its geometry with $G_p^0 \cong \mathrm{id}$ has to be maximal.

  The subgroup H is a discrete subgroup of G^0 and G^0 itself is unimodular. Milnor showed in [Mil76] that there are only six simply connected unimodular 3-dimensional Lie groups: $\mathbb{S}^3$, $\widetilde{\mathrm{SL}_2}(\mathbb{R})$, $\widetilde{\mathrm{Iso}(\mathbb{R}^2)}$, Sol, Nil and $\mathbb{R}^3$. So, by the maximal assumption we end up with the geometry $(\mathrm{Sol}, \mathrm{Iso}(\mathrm{Sol}))$. Compare Chapter 2.

□

Remark 2. Observe that the bundle curvature $\hat{\tau}$ defined by the curvature form in the proof is related to the bundle curvature τ defined in Chapter 1 for any Riemannian fibration by $\hat{\tau} = -2\tau$.

4 Geometric models for homogeneous manifolds

As we have seen, $E(\kappa,\tau)$ is a Riemannian fibration over a space-form $\Sigma(\kappa)$. Sometimes it is useful to consider coordinates: They exist at least for a local model for $E(\kappa,\tau)$. We endow $\mathbb{R}^3$ (for $\kappa \geqslant 0$) and $D\left(2/\sqrt{-\kappa}\right)\times\mathbb{R}$ (for $\kappa < 0$) with the metric

$$ds^2 = \lambda^2(dx_1^2 + dx_2^2) + (\tau\lambda(x_2\, dx_1 - x_1\, dx_2) + dx_3)^2,$$

where

$$\lambda = \frac{4}{4+\kappa(x_1^2+x_2^2)}.$$

For $\kappa > 0$ the manifold $(\mathbb{R}^3, ds^2)$ corresponds to the universal covering of $E(\kappa,\tau)$ minus one fibre. For $\kappa \leqslant 0$ the coordinates are global.

The Riemannian fibration $\pi\colon E(\kappa,\tau) \to \Sigma(\kappa)$ for these coordinates is given by the projection onto the first two coordinates. The vertical vector field is $\xi = \partial_{x_3}$.

To define an appropriate orthonormal frame $\{E_1, E_2, E_3\}$ of $TE(\kappa,\tau)$, we start with an orthonormal frame $\{e_1, e_2\}$ of $T\Sigma(\kappa)$, where

$$\Sigma(\kappa) = \begin{cases} \left(D\left(2/\sqrt{-\kappa}\right), \lambda^2(dx_1^2 + dx_2^2)\right), & \text{if } \kappa < 0, \\ (\mathbb{R}^2, \lambda^2(dx^2+dy^2)), & \text{if } \kappa \geqslant 0. \end{cases}$$

We define $e_1 = \lambda^{-1}\partial_{x_1}$, $e_2 = \lambda^{-1}\partial_{x_2}$; $E_i := \widetilde{e_i}$, $i = 1,2$ as their horizontal lifts:

$$\begin{aligned} E_1 &= \lambda^{-1}\partial_{x_1} - \tau x_2 \partial_{x_3}, \\ E_2 &= \lambda^{-1}\partial_{x_2} + \tau x_1 \partial_{x_3}, \quad \text{and} \\ E_3 &:= \partial_{x_3}. \end{aligned}$$

Later in Chapter 6 we derive the non-parametric mean curvature equation for surfaces in $E(\kappa,\tau)$. For this we need the Riemannian connection ∇ on $E(\kappa,\tau)$. The Levi-Civita connections ∇^Σ and ∇ on the base Σ and total space E of a Riemannian fibration are related by

$$\nabla_{\widetilde{X}}\widetilde{Y} - \widetilde{\nabla^\Sigma_X Y} = \frac{1}{2}\left[\widetilde{X},\widetilde{Y}\right]^v.$$

In our case, we compute for example

$$\nabla_{E_1} E_1 = \widetilde{\nabla^{\Sigma}_{e_1} e_1} \qquad \text{and}$$
$$\nabla^{\Sigma}_{e_1} e_1 = \lambda^{-1}(\partial_{x_1}\lambda^{-1}\partial_{x_1} + \lambda^{-1}\nabla^{\Sigma}_{\partial_{x_1}}\partial_{x_1})$$
$$= \lambda^{-1}\left(\frac{\kappa x_1}{2}\partial_{x_1} + \lambda^{-1}(\Gamma^1_{11}\partial_{x_1} + \Gamma^2_{11}\partial_{x_2})\right).$$

Where Γ^k_{ij} are the Christoffel symbols of Σ, therefore we determine them at first

$$\Gamma^1_{11} = \Gamma^2_{12} = \Gamma^2_{21} = -\Gamma^1_{22} = -\frac{\kappa\lambda}{2}x_1$$
$$\Gamma^2_{22} = \Gamma^1_{12} = \Gamma^1_{21} = -\Gamma^2_{11} = -\frac{\kappa\lambda}{2}x_2.$$

Using this we have

$$\nabla_{E_1} E_1 = \widetilde{\frac{\kappa x_2}{2\lambda}\partial_{x_2}} = \frac{\kappa x_2}{2}E_2.$$

Equivalent we compute

$$\nabla_{E_2} E_2 = \frac{\kappa x_1}{2}E_1.$$

Since E_3 is a Killing field we have $\langle \nabla_X E_3, Y\rangle + \langle X, \nabla_Y E_3\rangle = 0$ for all vector fields X, Y, furthermore with $X = E_3$ unitary we derive $\langle \nabla_{E_3} E_3, Y\rangle = -\langle E_3, \nabla_Y E_3\rangle = -1/2\partial_Y\langle E_3, E_3\rangle = 0$. So we conclude

$$\nabla_{E_3} E_3 = 0.$$

From $\partial_{E_i}\langle E_i, E_j\rangle = 0 = \partial_{E_j}\langle E_i, E_i\rangle$ for $i, j \in \{1,2,3\}$ follows $\langle \nabla_{E_i} E_i, E_j\rangle = -\langle \nabla_{E_i} E_j, E_i\rangle = 0$. With the Koszul formula we compute directly $\langle E_1, [E_1, E_2]\rangle = -\kappa x_2/2$ and $\langle E_2, [E_1, E_2]\rangle = \kappa x_1/2$. A short calculation shows

$$[E_1, E_2] = \frac{-\kappa x_2}{2}E_1 + \frac{\kappa x_1}{2}E_2 + \frac{\tau}{2\lambda}(4 - \lambda\kappa(x_1^2 + x_2^2))E_3.$$

Another straight computation shows $[E_i, E_3] = 0, i = 1,2,3$.

Using the Koszul formula again gives for example

$$2\langle \nabla_{E_1} E_2, E_i\rangle = \langle E_1, [E_i, E_2]\rangle - \langle E_2, [E_1, E_i]\rangle - \langle E_i, [E_2, E_1]\rangle$$

and therefore

$$\nabla_{E_1} E_2 = \frac{-\kappa x_2}{2}E_1 + \frac{\tau}{4\lambda}(4 - \lambda\kappa(x_1^2 + x_2^2))E_3.$$

We compute the other derivatives accordingly. The connection is given by

$$
\begin{array}{lll}
\nabla_{E_1} E_1 = \frac{\kappa x_2}{2} E_2, & \nabla_{E_2} E_1 = \frac{-\kappa x_1}{2} E_2 - \sigma E_3, & \nabla_{E_3} E_1 = -\sigma E_2, \\
\nabla_{E_1} E_2 = \frac{-\kappa x_2}{2} E_1 + \sigma E_3, & \nabla_{E_2} E_2 = \frac{\kappa x_1}{2} E_1, & \nabla_{E_3} E_2 = \sigma E_1, \\
\nabla_{E_1} E_3 = -\sigma E_2, & \nabla_{E_2} E_3 = \sigma E_1, & \nabla_{E_3} E_3 = 0,
\end{array}
$$

where

$$
\sigma := \frac{\tau}{4\lambda}(4 - \lambda\kappa(x_1^2 + x_2^2)).
$$

Part II.

Minimal and constant mean curvature surface theory

5 Hypersurfaces in Riemannian manifolds

Let us consider a hypersurface $f : M \to \bar{M}$, where M is a n-dimensional manifold and $(\bar{M}, \bar{g})$ a Riemannian manifold. If we endow M with the induced Riemannian metric g defined by

$$g(X,Y) := \bar{g}(\mathrm{d}f(X), \mathrm{d}f(Y)),$$

then the surface is *isometrically immersed*. Furthermore, if the hypersurface is orientable, there exists a global unit normal field $N : M \to \mathrm{T}\bar{M}$.

Definition 5.1. 1. The *shape operator* of (f,N) is the map

$$S\colon \mathrm{T}M \to \mathrm{T}M,\, S(X) = -\mathrm{d}f^{-1}\left(\bar{\nabla}_{\mathrm{d}f(X)}N\right),$$

where $\bar{\nabla}$ is the Riemannian connection of $\bar{M}$.

2. The *second fundamental form* of (f,N) is the symmetric bilinear form

$$b\colon \mathrm{T}M \times \mathrm{T}M \to \mathbb{R},\ b(X,Y) = g(SX,Y).$$

3. The *mean curvature* of (f,N) at $p \in M$ is

$$H(p) := \frac{1}{n}\operatorname{tr} S_p,$$

A *minimal surface* is a surface with $H \equiv 0$.

4. The *mean curvature vector* of f is the normal vector

$$\boldsymbol{H}(p) = H(p)N(p)$$

and does not depend on the chosen unit normal N.

5. A Riemannian submanifold M of a Riemannian manifold $\bar{M}$ is called *totally geodesic* if all geodesics in M are also geodesics in $\bar{M}$.

For 3-dimensional space-forms it is well-known that the *Gauß and Codazzi equations* for the first and second fundamental forms are the sufficient and necessary integrability conditions for the existence of a surface with these forms; the so-called *fundamental theorem of surfaces.* In [Dan07] Daniel proved the necessary and sufficient conditions to immerse a surface isometrically into a homogeneous manifold $E(\kappa, \tau)$.

Theorem 5.2 ([Dan07]). *Let $(M, \langle\cdot,\cdot\rangle)$ be a simply connected 2-dimensional Riemannian manifold and ∇ its covariant derivative. Let S be a field of symmetric operators $\mathrm{S}_y \colon \mathrm{T}_y M \to \mathrm{T}_y M$, T a vector field and ν a smooth function on M, such that $\|T\|^2 + \nu^2 = 1$.*

Let $E(\kappa, \tau)$ be a simply connected Riemannian fibration with geodesic fibres over $\Sigma^2(\kappa)$, bundle curvature τ and ξ a vector field tangent to the fibres.

Then there exists an isometric immersion $f \colon (M, \langle\cdot,\cdot\rangle) \to E(\kappa, \tau)$ such that the shape operator of (f, N) is S and $\xi = \mathrm{d}f(T) + \nu N$ if and only if for all vector fields X and Y on M the Gauß and Codazzi equations

$$K = \det \mathrm{S} + \tau^2 + (\kappa - 4\tau^2)\nu^2, \tag{5.1}$$

$$\nabla_X \mathrm{S}Y - \nabla_Y \mathrm{S}X - \mathrm{S}[X,Y] = (\kappa - 4\tau^2)\nu(\langle Y, T\rangle X - \langle X, T\rangle Y), \tag{5.2}$$

and

$$\nabla_X T = \nu(\mathrm{S}X - \tau JX), \tag{5.3}$$

$$\partial_X \nu = -\langle \mathrm{S}X - \tau JX, T\rangle. \tag{5.4}$$

are satisfied, where K is the Gauß curvature of M and J the $\pi/2$ rotation on $\mathrm{T}M$.

Furthermore, the immersion is unique up to any isometry of $E(\kappa, \tau)$ that preserves the orientations of the fibres and the base of the fibration.

Remark 3. Equations (5.1) - (5.4) are called *compatibility conditions* for $E(\kappa, \tau)$.

6 Constant mean curvature surfaces

We generalise the notion of a graph of a function to study one class of surfaces in Riemannian fibrations with geodesic fibres.

Definition 6.1. Let $\pi\colon E \to B$ be a fibre bundle over a base space B. Then a continuous map $s\colon B \to E$ is called *section* if $\pi(s(x)) = x$ for all $x \in B$.

Let E be the Riemannian fibration $E(\kappa,\tau)$. We call the surface $\{s(x) \in E(\kappa,\tau)\colon x \in \Omega\}$ a *graph* over $\Omega \subset \Sigma^2(\kappa)$ if s is transversal to the fibres.

Let M be a coordinate graph $z = u(x,y)$ in $E := E(\kappa,\tau)$ endowed with the metric from Chapter 4. With respect to the orthonormal frame (E_1, E_2, E_3) given therein, the tangent vector fields of the surface $f(x,y) = (x,y,u(x,y))$ are

$$\begin{aligned} f_x(x,y) &= \lambda E_1 + (u_x + \lambda\tau y)E_3 \\ f_y(x,y) &= \lambda E_2 + (u_y - \lambda\tau x)E_3, \end{aligned}$$

where $\lambda(x,y) = 4/(4 + \kappa(x^2 + y^2))$. Define $U := u_x + \lambda\tau y$ and $V := u_y - \lambda\tau x$. With this notation we compute the first fundamental form

$$g = \begin{pmatrix} \lambda^2 + U^2 & UV \\ UV & \lambda^2 + V^2 \end{pmatrix}.$$

For the upper normal N set $W := \sqrt{1 + U^2 + V^2}$. A straightforward computation gives

$$N = \frac{-UE_1 - VE_2 + \lambda E_3}{W}.$$

To compute the mean curvature we consider $T_1, T_2 \in \mathrm{T}M$ such that $(\mathrm{d}f(T_1), \mathrm{d}f(T_2), N)$ is an orthonormal frame of $\mathrm{T}E$. Since N has unit length, we have $\langle \nabla^E_N N, N\rangle = 0$ and therefore

$$\begin{aligned} 2H = \operatorname{tr} S_N &= -\sum_{i=1}^{2} \langle \nabla^E_{T_i} N, T_i\rangle \\ &= -\left(\sum_{i=1}^{2} \langle \nabla^E_{T_i} N, T_i\rangle + \langle \nabla^E_N N, N\rangle\right) \\ &= -\operatorname{div}_E(N). \end{aligned}$$

From the orthonormal frame (E_1, E_2, E_3) we get

$$\begin{aligned}
\operatorname{div}_E(N) &= \sum_{i=1}^{3} \langle \nabla^E_{E_i} N, E_i \rangle_E \\
&= \sum_{i=1}^{2} \langle \nabla^E_{E_i} N, E_i \rangle_E + \underbrace{\langle \nabla^E_{E_3} N, E_3 \rangle_E}_{=\partial_{E_3}(\lambda/W) - \langle N, \nabla_{E_3} E_3 \rangle = 0} \\
&\overset{(*)}{=} \sum_{i=1}^{2} \langle \nabla^E_{E_i} \left(\frac{(-UE_1 - VE_2}{W} \right), E_i \rangle_E \\
&= \sum_{i=1}^{2} \langle \nabla^\Omega_{e_i} \mathrm{d}\pi \left(\frac{-UE_1 - VE_2}{W} \right), e_i \rangle_\Omega \\
&= \sum_{i=1}^{2} \langle \nabla^\Omega_{e_i} \left(\frac{-Ue_1 - Ve_2}{W} \right), e_i \rangle_\Omega \\
&= \operatorname{div}_{\Sigma^2(\kappa)} \left(\frac{-Ue_1 - Ve_2}{W} \right),
\end{aligned}$$

in $(*)$ we used the fact that $\langle \nabla^E_{E_i} \frac{\lambda}{W} E_3, E_i \rangle = \langle E_i(\frac{\lambda}{W}) E_3, E_i \rangle + \langle \frac{\lambda}{W} \nabla_{E_i} E_3, E_i \rangle = 0$. We have $e_1 = \lambda^{-1}\partial_x$ and $e_2 = \lambda^{-1}\partial_y$ therefore, we may express H in terms of ∂_x and ∂_y:

$$\operatorname{div}_E(N) = \operatorname{div}_{\Sigma^2(\kappa)} \left(\frac{-U\partial_x - V\partial_y}{\lambda W} \right) = -2H$$

and derive the *non-parametric mean curvature surface equation*:

$$\begin{aligned}
2H &= \operatorname{div}_{\Sigma^2(\kappa)} \left(\frac{U\partial_x + V\partial_y}{\lambda W} \right) \\
&= \frac{1}{\sqrt{\det g_{\Sigma^2(\kappa)}}} \left(\partial_x \left(\sqrt{\det g_{\Sigma^2(\kappa)}} \frac{U}{\lambda W} \right) + \partial_y \left(\sqrt{\det g_{\Sigma^2(\kappa)}} \frac{V}{\lambda W} \right) \right) \\
&= \frac{1}{\lambda^2} \operatorname{div}_{\mathbb{R}^2} \begin{pmatrix} \frac{\lambda U}{W} \\ \frac{\lambda V}{W} \end{pmatrix}.
\end{aligned}$$

We proved the following theorem:

Theorem 6.2. *Let M be a coordinate graph $z = u(x,y)$ in $E(\kappa,\tau)$ with upward pointing normal N. Then the mean curvature $H(x,y,u(x,y)) = H$ satisfies*

$$H = \frac{1}{2\lambda^2}\operatorname{div}_{\mathbb{R}^2}\begin{pmatrix} \frac{\lambda U}{W} \\ \frac{\lambda V}{W} \end{pmatrix},$$

where $U = u_x + \lambda\tau y$, $V = u_y - \lambda\tau x$, and $W = \sqrt{1+U^2+V^2}$.

A powerful tool in surface theory is the maximum principle: Geometrically speaking, it states that two hypersurfaces under certain mean curvature conditions cannot touch each other one-sided. The maximum principle for surfaces follows from the maximum principle for elliptic partial differential equations due to Hopf.

Theorem 6.3. *Let M_1, $M_2 \subset E = E(\kappa,\tau)$ be two surfaces with constant mean curvature H and $\nu_1(p) = \nu_2(p)$ for $p \in S_1 \cap S_2$. If the signed distance function d_1 of S_1 in a neighbourhood $U = U_p \subset E$ does not change sign on $S_2 \cap U$, i.e. $d_1|_{S_2 \cap U} \geqslant 0$ or $d_1|_{S_2 \cap U} \leqslant 0$, then $S_1 \cap U = S_2 \cap U$.*

Remark 4. The maximum principle is also true in a more general setting, for a geometric discussion see [Esc89].

We finish this chapter with the definition of a more general class of surfaces. We want to construct surfaces with *finite topology*, i.e. surfaces that are homeomorphic to a compact surface with genus g, where k points are removed. We call the neighbourhood of such a point an *end*. In CMC surface theory one cannot always expect those surfaces to be embedded, therefore we consider the class of *Alexandrov-embedded* surfaces.

Definition 6.4 ([GKS03]). A CMC surface M of finite topology is Alexandrov-embedded if M is properly immersed, each end of M is embedded, and there exists a compact three-manifold W with boundary $\partial W =: \Sigma$ and a proper immersion $F\colon W\backslash\{q_1,\dots,q_k\} \to \bar{M}$ whose boundary restriction $f : \Sigma\backslash\{q_1,\dots,q_k\} \to \bar{M}$ parametrizes M. Moreover, we require that the mean-curvature normal of M points into W.

Remark 5. A non Alexandrov-embedded surface is, for example, a Delaunay nodoid. One can construct a triundoloid that is Alexandrov-embedded, but not embedded.

7 Sister surfaces

Lawson made a great contribution in the study of constant mean curvature surfaces in 1970 when he showed the correspondence between minimal surfaces in $\mathbb{S}^3$ and H-surfaces in $\mathbb{R}^3$. By means of the reflection principle in $\mathbb{S}^3$ it was then possible to construct new surfaces in the Euclidean space. In recent years mathematicians have grown their interest within surfaces in other ambient manifolds. Nelli and Rosenberg showed some examples in $\mathbb{H}^2 \times \mathbb{R}$ ([NR06]). In 2007 Daniel published a generalized Lawson correspondence for homogeneous manifolds ([Dan07]), we use one special case of his correspondence:

Theorem 7.1 ([Dan07, Theorem 5.2]). *There exists an isometric correspondence between an MC H-surface $\widetilde{M}$ in $\Sigma(\kappa) \times \mathbb{R} = E(\kappa, 0)$ and a minimal surface M in $E(\kappa + 4H^2, H)$. Their shape operators are related by*

$$\widetilde{S} = JS + H\,\mathrm{id}, \tag{7.1}$$

where J denotes the $\pi/2$ rotation on the tangent bundle of a surface. Moreover, the normal and tangential projections of the vertical vector fields ξ and $\widetilde{\xi}$ are related by

$$\langle \widetilde{\xi}, \widetilde{\nu} \rangle = \langle \xi, \nu \rangle, \qquad J\,\mathrm{d}f^{-1}(T) = \mathrm{d}\widetilde{f}^{-1}(\widetilde{T}), \tag{7.2}$$

where f and $\widetilde{f}$ denote the parametrizations of M and $\widetilde{M}$ respectively, ν and $\widetilde{\nu}$ their unit normals, and T, $\widetilde{T}$ the projections of the vertical vector fields on $\mathrm{T}M$ and $\mathrm{T}\widetilde{M}$.

We call the isometric surfaces M and $\widetilde{M}$ *sister surfaces*, or *sisters* in short.
Examples:

1. For $H = 0$ the surfaces M and $\widetilde{M}$ are conjugate minimal surfaces in $\Sigma(\kappa) \times \mathbb{R}$.
2. For $H \in (0, 1/2)$ and $\kappa = -1$ we have $4H^2 - 1 < 0$ and therefore corresponds to a minimal surface in $\widetilde{\mathrm{PSL}_2}(\mathbb{R})$ and its MC H-sister surface in $\mathbb{H}^2 \times \mathbb{R}$.
3. For $H = 1/2$ and $\kappa = -1$ an MC 1/2-surface in $\mathbb{H}^2 \times \mathbb{R}$ corresponds to a minimal surface in $E(0, 1/2) = \mathrm{Nil}_3(\mathbb{R})$.
4. Furthermore, for $H > 1/2$ an MC H-surface in $\mathbb{H}^2 \times \mathbb{R}$ results from a minimal surface in the Berger spheres $E(4H^2 - 1, H)$, since $4H^2 - 1 > 0$.

7.1 Reflection principles

We want to apply Schwarz reflection to construct complete periodic CMC surfaces in $E(\kappa,\tau)$. It is well-known that the Schwarz reflection extends minimal surfaces in space forms (see [Law70]) with respect to the space form symmetries. In $E(\kappa,\tau)$ the isometry group has dimension 4 only, but still there are symmetries, for which the Schwarz reflection applies.

Definition 7.2. Let $c \subset E(\kappa,\tau)$ be a geodesic and $V \subset E(\kappa,\tau)$ be a totally geodesic plane.

1. A *geodesic reflection* across c is a map $\sigma_c\colon E(\kappa,\tau) \to E(\kappa,\tau)$ such that

$$\sigma_c(\gamma(s)) = \gamma(-s), \text{ for all } s < |r|,$$

for any arclength-parametrized geodesic $\gamma\colon (-r,r) \to E(\kappa,\tau)$ perpendicular to c with $\gamma(0) \in c$.

2. A *geodesic reflection* across V is a map $\sigma_V\colon E(\kappa,\tau) \to E(\kappa,\tau)$ such that

$$\sigma_V(\gamma(s)) = \gamma(-s), \text{ for all } s < |r|,$$

for any arclength-parametrized geodesic $\gamma\colon (-r,r) \to E(\kappa,\tau)$ perpendicular to V with $\gamma(0) \in V$.

Remark 6. The geometric interpretation of the reflection is to send a point p to its opposite point on a geodesic through p that meets c (or V) orthogonally.

Lemma 7.3 ([GK10]). *1. If c is a horizontal or vertical geodesic of $E(\kappa,\tau)$, then the geodesic reflection across c is an isometry.*

2. In the product spaces $E(\kappa,0) = \Sigma(\kappa) \times \mathbb{R}$ a geodesic reflection across vertical or horizontal planes is an isometry.

Proof. 1. The projection $\pi(c)$ of a horizontal geodesic is a geodesic in $\Sigma(\kappa)$. In $\Sigma(\kappa)$ the reflection across a geodesic is an isometry. Since $E(\kappa,\tau)$ is a Riemannian fibration and the composition of isometries is an isometry, σ_c is an isometry.

For a vertical geodesic c, the map σ_c can be expressed as the composition of two reflections across horizontal geodesics.

2. In the product spaces vertical and horizontal planes are totally geodesic. Hence, by the first part of the lemma, the map σ_V is an isometry.

□

Proposition 7.4 (Reflection principles, [GK10]). *1. Suppose that a minimal surface is smooth up to the boundary and the boundary contains a curve which is a horizontal or vertical geodesic of $E(\kappa,\tau)$. Then the symmetry of Lemma 7.3 extends the surface smoothly.*

2. The same holds for a CMC surface in a product space $E(\kappa,0)$, which is smooth up to the boundary, in case the boundary contains a curve in a vertical or horizontal plane, provided the surface conormal is perpendicular to the plane.

A totally geodesic plane is called *mirror plane*, and a curve in which the surface meets a mirror plane orthogonally is called *mirror curve*.

Remark 7. We construct surfaces by solving Plateau problems. The solution is an area minimising map from a disc to $E(\kappa,\tau)$, continuous up to the boundary. To apply Proposition 7.4 we have to exclude branch points in the interior ([Oss70] and [Gul73]) and at the boundary (see Proposition 8.4).

7.2 Sister curves

We want to analyse the geometry of periodic surfaces. Therefore, we take a closer look at their boundary curves.

Definition 7.5. Let $c = f \circ \gamma$ be a curve parametrized by arc length in a hypersurface $M = f(\Omega) \subset \bar{M}$ with (surface) normal ν. The *normal curvature* k and the *normal torsion* t along c are defined by

$$k := \nu \cdot \bar{\nabla}_{c'} c' = -\bar{\nabla}_{c'} \nu \cdot c' = \langle S\gamma', \gamma' \rangle, \qquad t := -\bar{\nabla}_{c'} \nu \cdot Jc' = \langle S\gamma', J\gamma' \rangle.$$

Let $\widetilde{M} \subset \Sigma(\kappa) \times \mathbb{R}$ denote an MC H-surface and $M \subset E(\kappa + 4H^2, H)$ its minimal sister. Furthermore, let γ be a curve in Ω. We call $\widetilde{c} := \widetilde{f}(\gamma)$ and $c := f(\gamma)$ *sister curves*.

Lemma 7.6. *For a pair of sister curves the normal curvature and torsion are related as follows:*

$$\widetilde{k} = -t + H \quad \text{and} \quad \widetilde{t} = k.$$

Proof. We apply Equation (7.1) to the definitions:

$$\widetilde{k} = \langle \widetilde{S}\gamma', \gamma' \rangle = \langle (JS + H\,\mathrm{id})\gamma', \gamma' \rangle = -t + H,$$
$$\widetilde{t} = \langle \widetilde{S}\gamma', J\gamma' \rangle = \langle (JS + H\,\mathrm{id})\gamma', J\gamma' \rangle = k. \qquad \square$$

Now we are able to express the relation between mirror curves and their sister curves.

Proposition 7.7 ([GK10],[MT11]). 1. *A curve $\widetilde{c} \subset \widetilde{M} \subset \Sigma(\kappa) \times \mathbb{R}$ is a vertical mirror curve if and only if its sister curve $c \subset M \subset E(\kappa+4H^2, H)$ is a horizontal geodesic.*

2. *Similarly, $\widetilde{c}$ is a horizontal mirror curve if and only if c is a vertical geodesic.*

7.3 Geometric quantities

In our construction we consider as in [Gro93] the *fundamental patch* of a periodic MC H-surface. The complete surface is then generated by reflections. The fundamental patch is simply connected and bounded by mirror curves $\widetilde{c}_i$. Given a fundamental patch bounded by n arc length parametrized mirror curves $\widetilde{c}_i$, it defines the following geometric quantities:

1. The *length* $\widetilde{l}_i$ of the mirror curve $\widetilde{c}_i$, also denoted by $l(\widetilde{c}_i)$ or $|\widetilde{c}_i|$.

2. The *vertex angle* $\widetilde{\varphi}_i$ of two edges $\widetilde{c}_i$ and $\widetilde{c}_{i+1}$, which satisfies

$$\cos\widetilde{\varphi}_i = -\widetilde{c}_i'(\widetilde{l}_i) \cdot \widetilde{c}_{i+1}'(0).$$

3. The total *turn* angle of the normal $\widetilde{\nu}$

$$\mathrm{turn}_i = \mathrm{turn}_{\widetilde{c}_i}(\widetilde{\nu}) := \int_{\widetilde{c}_i} \widetilde{k},$$

which measures the total turn of the normal relative to a parallel field.

Proof of the interpretation of the turn: Let N be a Riemannian manifold and X, Y be two unit vector fields along a curve $c \subset N$, such that $\mathrm{span}\{X, Y\} \subset TN$ is a two-dimensional subspace. Let J denote the $\pi/2$ rotation in $\mathrm{span}\{X, Y\}$, then $(Y, JY, Y \times JY)$ is a positively oriented frame of TN. If $\alpha(t)$ measures the angle $\measuredangle(X(t), Y(t))$, then $X = \cos\alpha Y + \sin\alpha JY$ and $JX = \cos\alpha JY - \sin\alpha Y$.

We look at the covariant derivative

$$\begin{aligned}\nabla^E_{\tilde{c}'}X &= \nabla^E_{\tilde{c}'}(\cos\alpha Y + \sin\alpha JY)\\ &= -\alpha'\sin\alpha Y + \cos\alpha\nabla^E_{\tilde{c}'}Y + \alpha'\cos\alpha JY + \sin\alpha\nabla^E_{\tilde{c}'}JY.\end{aligned}$$

The product with JX is

$$\begin{aligned}&\langle\nabla^E_{\tilde{c}'}X, JX\rangle_E =\\ &\quad\langle -\alpha'\sin\alpha Y + \cos\alpha\nabla^E_{\tilde{c}'}Y + \alpha'\cos\alpha JY + \sin\alpha\nabla^E_{\tilde{c}'}JY, \cos\alpha JY - \sin\alpha Y\rangle\\ &\quad= \cos^2\alpha\langle\nabla^E_{\tilde{c}'}Y, JY\rangle + \alpha'\cos^2\alpha + \cos\alpha\sin\alpha\underbrace{\langle\nabla^E_{\tilde{c}'}JY, JY\rangle}_{=0,\ \text{since}\ \langle JY,JY\rangle=1}\\ &\qquad - \cos\alpha\sin\alpha\underbrace{\langle\nabla^E_{\tilde{c}'}Y, Y\rangle}_{=0} + \alpha'\sin^2\alpha - \sin^2\alpha\underbrace{\langle\nabla^E_{\tilde{c}'}JY, Y\rangle}_{=-\langle\nabla^E_{\tilde{c}'}Y,JY\rangle}\\ &\quad= \alpha' + \langle\nabla^E_{\tilde{c}'}Y, JY\rangle,\end{aligned}$$

hence, we get

$$\alpha'(t) = \langle\nabla^E_{\tilde{c}'}X, JX\rangle - \langle\nabla^E_{\tilde{c}'}Y, JY\rangle.$$

In our case X, Y in TV, where V is a mirror plane in $E(\kappa, 0)$ and M is an MC H-surface with normal ν that meets V orthogonally. With $X = \nu$ and parallel Y ($\nabla^E_{\tilde{c}'}Y = 0$) we get

$$\alpha' = \langle\nabla^E_{\tilde{c}'}\nu, -\tilde{c}'\rangle = \tilde{k}.$$

□

By Proposition 7.7 the minimal sister surface is bounded by horizontal and vertical geodesics. Since the surfaces are isometric, we have

$$\tilde{l}_i = l_i, \qquad \tilde{\varphi}_i = \varphi_i. \tag{7.3}$$

Accordingly, we want to measure the rotational angle of the normal ν along a curve c in $E(\kappa + 4H^2, H)$. We detect the *twist* of the normal with respect to an appropiate vector field X

$$\operatorname{twist}_c(\nu, X) := \int_c \langle\nabla_{c'}\nu, c' \times \nu\rangle - \langle\nabla_{c'}X, c' \times X\rangle.$$

Definition 7.8. 1. Let $c \subset M$ be a vertical geodesic in $E(\kappa+4H^2,H)$, then the twist is defined by the total rotation speed of v with respect to a basic vector field $\tilde{e}$, i.e. the horizontal lift of any vector field e on $\Sigma(\kappa+4H^2)$:

$$\operatorname{twist}_v := \operatorname{twist}_c(v,\tilde{e}) = \int_c \langle \nabla_{c'} v, c' \times v\rangle - \langle \nabla_{c'} \tilde{e}, c' \times \tilde{e}\rangle.$$

2. Let $c \subset M$ be a horizontal geodesic in $E(\kappa+4H^2,H)$, then the twist is defined by the total rotation speed of v with respect to the vertical vector field ξ:

$$\operatorname{twist}_h := \operatorname{twist}_c(v,\xi) = \int_c \langle \nabla_{c'} v, c' \times v\rangle - \langle \nabla_{c'} \xi, c' \times \xi\rangle.$$

With this definition twist_v measures the angle $\measuredangle(d\pi_c(v(0)), d\pi_c(v(l))$ in the projection. Please note that the Definition 7.81 differs from the one in [GK10]. We drop the index when it is clear whether the geodesic is vertical or horizontal.

Lemma 7.9. *1. Let $c \subset M$ be a vertical geodesic in $E(\kappa+4H^2,H)$, whose sister is a horizontal mirror curve $\tilde{c} \subset \tilde{M}$ in $E(\kappa,0)$. Then*

$$\operatorname{twist}_v = \int_c t + Hl(c) \quad \text{and} \quad \tilde{k} = 2H - \operatorname{twist}'_v .$$

2. Let $c \subset M$ be a horizontal geodesic in $E(\kappa+4H^2,H)$, whose sister is a vertical mirror curve $\tilde{c} \subset \tilde{M}$ in $E(\kappa,0)$. Then

$$\operatorname{twist}_h = -\operatorname{turn}.$$

Proof. 1. Without loss of generality $c' = \xi$. Let (c', Jc', v) be positively oriented. Let J and R denote $\pi/2$-rotations in the tangent bundle TM and the horizontal plane $d\pi^{-1}(T\Sigma(\kappa+4H^2))$, respectively. Then Jc' and v are horizontal with $-Rv = Jc'$.

We denote by E an unit basic vector field. We have

$$\begin{aligned}
\operatorname{twist}_v &= \int_c (\langle \nabla_{c'} v, c' \times v\rangle - \langle \nabla_{c'} E, c' \times E\rangle) \\
&= \int_c (\langle \nabla_{c'} v, Rv\rangle - \langle \nabla_{c'} E, RE\rangle) \\
&= \int_c (-\langle \nabla_{c'} v, Jc'\rangle - \langle \nabla_{\xi} E, RE\rangle) \\
&= \int_c (t + H\langle RE, RE\rangle) \\
&= \int_c t + Hl(c).
\end{aligned}$$

Lemma 7.6 implies that $\widetilde{k} = 2H - \operatorname{twist}'_v$.

2. The rotational angle of the tangent plane $\mathrm{T}_{c(t)} M$ is measured by the rotational speed with respect to ξ:

$$\operatorname{twist}'_h = \langle \nabla_{c'} v, \underbrace{c' \times v}_{=-Jc'}\rangle - \langle \underbrace{\nabla_{c'}\xi}_{=-HRc'}, \underbrace{c' \times \xi}_{=-Rc'}\rangle = t - H = -\widetilde{k}.$$

By integrating along c we get

$$\operatorname{twist}_h = -\operatorname{turn}. \quad \square$$

Example. We compute the torsion and the twist of a vertical geodesic c in a vertical plane in $E(\kappa, \tau)$. The geodesic c is a fibre of the Riemannian fibration. Without loss of generality $c' = \xi$, then we get

$$\begin{aligned}
t &= -\nabla_{c'} v \cdot Jc' \\
&= \tau R v \cdot Jc' = -\tau.
\end{aligned}$$

Namely, the torsion of a vertical geodesic is the negative of the bundle curvature.

Moreover, for the twist we get, as expected

$$\operatorname{twist} = \int_c t + \tau l(c) = 0.$$

Hence, with respect to parallel fields, the normal does not rotate. Equivalently the normal is constant in the projection.

We shall apply Lemma 7.9 to obtain detailed information about vertical geodesics in Nil and their horizontal sister curves in $\mathbb{H}^2 \times \mathbb{R}$. This strategy is due to Laurent Mazet, for whom the author is very grateful.

For a curve $\tilde{c} \subset \mathbb{H}^2$, consider the unique horocycle foliation $\mathscr{F}_{\tilde{c}}$ given by the horocycle that is tangent to $\tilde{c}$ in $\tilde{c}(0)$ and has curvature 1 with respect to the normal n of $\tilde{c}$. Let ϑ be the angle defined by $\tilde{c}' = \cos\vartheta e_1 - \sin\vartheta e_2$, where the orthonormal frame (e_1, e_2) is given by the tangent and the normal of the horocycles. By the definition of the considered foliation we have $\vartheta(0) = 0$ and $n = \sin\vartheta e_1 + \cos\vartheta e_2$.

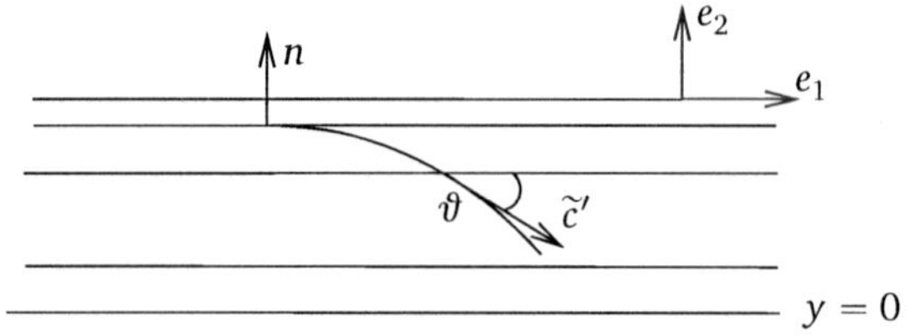

Figure 7.1.: Horocycle foliation $\mathscr{F}_{\tilde{c}}$ given by $\tilde{c}$.

Lemma 7.10. *The curvature of $\tilde{c}$ is given by*

$$\tilde{k} = \cos\vartheta - \vartheta'.$$

Proof. A simple computation gives:

$$\begin{aligned}
\nabla_{\tilde{c}'}\tilde{c}' =& \nabla_{\cos\vartheta e_1 - \sin\vartheta e_2}(\cos\vartheta e_1 - \sin\vartheta e_2) \\
=& -\vartheta' \sin\vartheta e_1 + \cos\vartheta \nabla_{(\cos\vartheta e_1 - \sin\vartheta e_2)} e_1 - \vartheta' \cos\vartheta e_2 - \sin\vartheta \nabla_{(\cos\vartheta e_1 - \sin\vartheta e_2)} e_2 \\
=& -\vartheta' \sin\vartheta e_1 + \cos^2\vartheta \underbrace{\nabla_{e_1} e_1}_{e_2} + \cos\vartheta \sin\vartheta \underbrace{\nabla_{e_2} e_1}_{=0} \\
& - \vartheta' \cos\vartheta e_2 - \sin\vartheta\cos\vartheta \underbrace{\nabla_{e_1} e_2}_{-e_1} + \sin^2\vartheta \underbrace{\nabla_{e_2} e_2}_{0} \\
=& -\vartheta' \sin\vartheta e_1 + \cos^2\vartheta e_2 - \vartheta' \cos\vartheta e_2 + \sin\vartheta\cos\vartheta e_1 \\
=& (-\vartheta' + \cos\vartheta)\sin\vartheta e_1 + (-\vartheta' + \cos\vartheta)(\cos\vartheta e_2) \\
=& (\cos\vartheta - \vartheta')n.
\end{aligned}$$

□

We want to control the curve $\widetilde{c}$ by means of its sister c in $M \subset E(0,1/2) = \mathrm{Nil}$. Let $\alpha(t) = \mathrm{twist}_v$ measure the twist in $c(t)$ with respect to a basic vector field chosen such that $\alpha(0) = 0$, i.e. $\widetilde{e}(c(0)) = \widetilde{v}(c(0))$.

Proposition 7.11. *Let $\widetilde{c} \subset \mathbb{H}^2 \times \mathbb{R}$ be the horizontal sister curve of a vertical geodesic c in $E(0,1/2) = \mathrm{Nil}$. The angle ϑ given by the horocycle foliation $\mathscr{F}_{\widetilde{c}}$ and the rotational speed α' of the minimal surface normal along c are related as follows:*

$$\vartheta' = \alpha' + \cos\vartheta - 1, \quad \textit{and} \quad \vartheta \leqslant \alpha,$$

where $\alpha(t)$ measures the angle between v and any parallel field chosen such that $\alpha(0) = 0$.

Proof. We have seen in Lemma 7.9 that $\widetilde{k} = 1 - \alpha'$. Together with Lemma 7.10 we get

$$\vartheta' = \alpha' + \cos\vartheta - 1,$$

namely $\vartheta' \leqslant \alpha'$. In particular, along the curves c and $\widetilde{c}$ we have

$$\int_{\widetilde{c}} \vartheta' \leqslant \int_c \alpha' \Rightarrow \vartheta(t) \leqslant \alpha(t).$$

□

8 Solution of the Plateau problem

From the 18th century until 1930 it was an open question whether a rectifiable closed curve Γ in $\mathbb{R}^3$ bounds an area minimizing surface. Douglas ([Dou31]) and Radó ([Rad30]) proved independently that there always exists a minimal surface of disc type spanned by Γ. In 1948 Morrey generalised the theorem to minimal surfaces in homogeneously regular Riemannian manifolds without boundary ([Mor66]). By [Oss70] and [Gul73], the least area disc is a minimal immersion in the interior. In the 80ties Meeks and Yau ([MY82]) showed that in manifolds with mean convex boundary the Plateau solution is even embedded.

Definition 8.1. A Riemannian manifold N with boundary is *mean convex* if the following conditions hold:

1. The boundary ∂N is piecewise smooth,
2. each smooth subsurface of ∂N has non-negative mean curvature with respect to the inward normal,
3. there exists a Riemannian manifold N' such that N is isometric to a submanifold of N' and each smooth subsurface S of ∂N extends to a smooth embedded surface S' in N' such that $S' \cap N = S$.

We call each surface S a *barrier*.

Theorem 8.2 ([HS88, Theorem 6.3]). *Let N be a compact Riemannian 3-manifold with mean convex boundary, and let Γ be a Jordan curve in ∂N which is null-homotopic in N. Then Γ bounds a least area disc which is either properly embedded in N or is embedded in ∂N.*

Under additional assumptions we may exclude boundary branch points, for instance when the boundaries are *special polygonal Jordan curves.*

Definition 8.3. A *special polygon* in a complete 3-manifold N is a piecewise geodesic Jordan curve Γ that has the following properties:

1. At each vertex, the angle is of the form π/n, with natural $n \geqslant 2$, and

2. for each edge there is an isometry of N which acts by π rotation about that edge.

Remark 8. We are interested in the case where $N = E(\kappa, \tau)$ and want to use the Schwarz reflection about the edges. For this we need to guarantee the existence of an isometry acting on the edges. For $p \in \Gamma$ we denote by $G_p \subset E(\kappa, \tau)$ the group of isometries generated by π rotations about those edges of Γ that contain p.

Proposition 8.4 ([GK10]). *Let M be a solution of the Plateau problem for a special polygonal Jordan curve Γ contained in the boundary of a mean convex domain Ω in some 3-manifold N. At each point $p \in \Gamma$, suppose that the images of Ω under the group G_p are disjoint. Then M is regular at the boundary in the sense that the extension of M across Γ by the Schwarz reflection group G_p yields a smoothly immersed surface.*

As stated above, in particular we consider surfaces that are sections, i.e. graphs in the context of Riemannian fibrations.

Lemma 8.5. *Let $\Delta := \pi(M)$ be a compact disc. If M does not have any vertical tangent planes, then M is a section over Δ.*

Proof. Since there is no vertical tangent plane we have $\mathrm{d}\pi v \neq 0$ for all $v \in \mathrm{T}_p M$ and any $p \in M$. By the inverse mapping theorem there exists a neighbourhood $U_p \subset \Delta$ and a continuous map $s_p \colon U_p \to E$ such that $\pi(s_p(x)) = x$ for all $x \in U_p$. Moreover, there exists a finite covering $\{U_{p_n}\}_n$ of Δ and the inverse maps coincide for $U_{p_k} \cap U_{p_l} \neq \emptyset$. Therefore, we get a continuous map $s : \Delta \to E$ such that $\pi(s(x)) = x$ for all $x \in \Delta$. □

In the case where $\pi(\partial M)$ bounds a proper subset $\Delta \subset \Sigma$ and $M \subset \pi^{-1}(\Delta)$, there is a generalized Radó lemma:

Proposition 8.6 ([GK10]). *Let $\pi : E \to \Sigma$ be a Riemannian fibration with geodesic fibres, and let $\Delta \subset \Sigma$ be a convex disc with non-empty boundary $\partial\Delta$. Suppose $M \subset \pi^{-1}(\Delta)$ is a compact, homotopically trivial minimal surface such that $\pi \colon \partial M \to \partial\Delta$ is injective except for at most finitely many points of $\partial\Delta$. Then M is a section over Δ.*

Like in $\mathbb{R}^3$, we can prove uniqueness:

Proposition 8.7. *Let $\pi \colon E \to \Sigma$ be a Riemannian fibration with geodesic fibres. Suppose M is a section over $\Delta \subset \Sigma$ with MC H and prescribed boundary values such that $\pi \colon \partial M \to \partial\Delta$ is injective. Then M is unique.*

Proof. Assume that we have two sections M and $\hat{M}$ over Δ with the same boundary values. Proposition 6.2 states, that they fulfil the non-parametric mean curvature equation. Therefore, their parametrizations u and $\hat{u}$ are solutions of the following differential equation:

$$Q(u) := \partial_x\left(\frac{\lambda U}{W}\right) + \partial_y\left(\frac{\lambda V}{W}\right) - 2\lambda^2 H = 0,$$

where $U = u_x + \lambda\tau y$, $V = u_y - \lambda\tau x$, $W = \sqrt{1+U^2+V^2}$ and $\lambda = \frac{4}{4+\kappa(x^2+y^2)}$.

This equation is non-linear. We set $a^i(p) := \frac{\lambda p_i}{\sqrt{1+p_1^2+p_2^2}}$ and $R := U\partial_x + V\partial_y$. Considering the difference $u - \hat{u}$ we get:

$$\begin{aligned}
0 = Q(u) - Q(\hat{u}) &= \sum_i \partial_i a^i(R) - \partial_i a^i(\hat{R}) \\
&= \sum_i \partial_i a^i(tR + (1-t)\hat{R})\big|_{t=0}^{t=1} \\
&= \int_0^1 \frac{\mathrm{d}}{\mathrm{d}s}\left[\sum_i \partial_i a^i(sR + (1-s)\hat{R})\right]_{s=t} \mathrm{d}t \\
&= \sum_{i,j}\int_0^1 \partial_i\left[\frac{\partial a^i}{\partial p_j}(tR+(1-t)\hat{R})(R_j - \hat{R}_j)\right]\mathrm{d}t \\
&= \sum_{i,j}\partial_i\left[\underbrace{\left(\int_0^1 \frac{\partial a^i}{\partial p_j}(tR+(1-t)\hat{R})\,\mathrm{d}t\right)}_{=:a^{ij}}(R_j - \hat{R}_j)\right] \\
&= \sum_{i,j}\partial_i[a^{ij}\partial_j(u-\hat{u})].
\end{aligned}$$

Since

$$\partial_j a^i(p) = \frac{\lambda\delta_{ij}}{\sqrt{1+p_1^2+p_2^2}} - \frac{\lambda p_i p_j}{\sqrt{1+p_1^2+p_2^2}^{\,3}} = \partial_i a^j(p),$$

then $(a^{ij})_{ij}$ is symmetric, moreover $\left|(\partial_i a^{ij})_j\right|$ is bounded. Therefore with $w := u - \hat{u}$, we have that

$$L(w) := \sum_{i,j}\partial_i(a^{ij}\partial_j w) = Q(u) - Q(\hat{u}) = 0$$

is a linear second order partial differential operator.

With $T(t,x,y) := tR(x,y)+(1-t)\hat{R}(x,y)$ we get $\partial_j a^i(T) = \frac{\lambda\delta_{ij}}{\sqrt{1+|T|^2}} - \frac{\lambda T_i T_j}{\sqrt{1+|T|^2}^3}$. For any compact subset $K \subset \Delta$ the norm $|T|$ has a maximum on $[0,1] \times K$. Hence, there exists $\sigma(K,u,\hat{u}) > 0$, such that we may estimate, by means of the Schwarz inequality,

$$\sum_{i,j} a^{ij}(T)\xi_i\xi_j = \frac{(1+|T|^2)\,|\xi|^2 - \langle T,\xi\rangle^2}{\sqrt{1+|T|^2}^3} \geqslant \frac{|\xi|^2}{\sqrt{1+|T|^2}^3} > \sigma\,|\xi|^2 .$$

We conclude that L is uniformly elliptic and therefore, the maximum principle for elliptic partial differential equations ([GT01]) applies. □

Remark 9. The uniqueness of a section is also true in a more general case: The projection of ∂M has to be injective except for at most finitely many points of $\partial\Delta$. This means, we allow vertical segments in the boundary. The proof needs a more general maximum principle by Nitsche; for $\mathbb{R}^3$ see [Nit75].

9 Convergence of minimal surfaces

We construct our examples as limits of a sequence of bounded minimal surfaces. To ensure convergence we need a locally uniform area bound and a curvature estimate.

Große-Brauckmann and Kusner proved a general curvature estimate for graphs in Riemannian fibrations with bounded geometry:

Proposition 9.1 ([GK10]). *Let $\pi\colon E(\kappa,\tau) \to \Sigma(\kappa)$ be a Riemannian fibration with geodesic fibres. Consider a family $\mathscr{M}$ of embedded minimal graphs of disc type $M \subset E$, whose (possibly empty) boundaries are special polygons of controlled geometry. Assume that for each $M \in \mathscr{M}$, there is a mean convex set $\Omega(M) \supset M$ with bounded geometry. Then there is a uniform curvature estimate for the family $\mathscr{M}$.*

The proof is by contradiction: Assume there is a sequence of points in the family $\mathscr{M}$, for which an absolute principal curvature goes to infinity. Then after a blow-up of the metric the ambient spaces subconverge to $\mathbb{R}^3$. Große-Brauckmann and Kusner showed that a subsequence of the minimal discs would converge to the plane. But this contradicts the Gauß curvature.

In the special case of a sequence of nested compact mean convex sets there exists a curvature estimate and therefore a limit.

Theorem 9.2 ([Gro93]). *Let N be a closed manifold with boundary. A sequence of Jordan curves $\Gamma_n \subset N$ bounds minimal discs that are sections M_n. If Γ_n converges to a non-closed curve Γ, then the sections M_n converge to M, where $\partial M = \Gamma$. The limit M is a minimal surface that is simply connected and has no interior branch points.*

Proposition 9.3. *Let M_n be a sequence of minimal discs, which are graphs in a Riemannian fibration $\pi\colon E(\kappa,\tau) \to \Sigma(\kappa)$ converging to a minimal disc M. Then M is either a graph over $\lim_{n\to\infty} \pi(M_n)$ or a vertical plane.*

Proof. Since each M_n is a graph, it is transversal to the fibres and for the normal ν_n we have $\langle \nu_n, \xi\rangle > 0$ for each $n \in \mathbb{N}$. For the limit $\nu_n \to \nu$ we still know that $\langle \nu, \xi\rangle \geqslant 0$. If $\langle \nu, \xi\rangle > 0$, then M is a graph. Otherwise we pick $p \in M$ with $\langle \nu(p), \xi\rangle = 0$ and consider the unique vertical plane $V \subset E(\kappa,\tau)$ that is orthogonal to ν in p. If $M \neq V$, we analyse $\Gamma := M \cap V$. By the maximum principle, Theorem 6.3, we have $\Gamma \neq \{p\}$.

Locally in a neighbourhood $U_p \subset V$ of p, the surface M is a graph M_p over V in the normal direction.

We claim: There exists a map $\hat{d}\colon M_p \subset M \to \mathbb{R}$ with $M_p \cap V = \{\hat{d} = 0\}$ such that its composition with a parametrisation f of M is a real analytic function $d := \hat{d} \circ f$.

We construct $\hat{d}$ via the diffeomorphism $e\colon M_p \times (-\delta, \delta) \to E(\kappa, \tau)$ defined by $e(q,t) = \exp^E(t\nu_M(q))$. The map $\hat{d} := \pi_{(-\delta,\delta)} \circ e^{-1}\colon M_p \to \mathbb{R}$ is the signed distance function of M. We have $M \cap U_p = \{\hat{d} = 0\}$.

If we choose a harmonic and conformal parametrisation $f\colon D \to E(\kappa, \tau)$ of M, where $D \subset \mathbb{R}^2$ is the open unit disc, then the function $d := \hat{d} \circ f$ is analytic. Therefore, it can be expressed as a power series $d(x) = \sum_{j=m}^{\infty} P_j(x)$, where P_j is a polynomial of degree j and $P_m \neq 0$. Since the first derivatives of V and M coincide, we have $m \geqslant 2$.

Moreover, Γ divides U_p in at least $2m$ regions. There exists $q \in M \backslash \Gamma$ with a neighbourhood M_q, such that $M_q \cap \Gamma = \emptyset$ and $\pi_{\Sigma(\kappa)}(M_q)$ is not injective. But since M is the limit of M_n this means that there exists $n \in \mathbb{N}$ for which $\pi_{\Sigma(\kappa)}(M_n)$ is not injective, a contradiction to M_n being a graph.

Therefore, $P_m = 0$ and $d \equiv 0$, which means that M is a vertical plane. □

The fact that the intersection of two non-transversally intersecting minimal surfaces consists of $2m$ embedded curves is true in general (see [CM11]).

Part III. Examples

10 Known surfaces

10.1 Ruled surfaces in homogeneous manifolds

In [GK09] Große-Brauckmann and Kusner discussed ruled minimal surfaces in homogeneous manifolds and their sisters systematically. Since we need some of them as barriers in our surface construction, we present a short outline of their work.

In $\mathbb{R}^3$ a ruled surface is defined for $I \subset \mathbb{R}$, $c\colon I \to \mathbb{R}^3$ and $v\colon I \to \mathbb{S}^2$ as the mapping

$$f\colon \mathbb{R} \times I \to \mathbb{R}^3, \quad f(s,t) := c(t) + sv(t).$$

It does not need to be an immersion, but this follows from minimality. The curve c is called directrix; the rulings $\gamma(s) := f(s,t_0)$ are asymptote lines, i.e. $\kappa_{\mathrm{norm}}(\gamma) := g(S\gamma',\gamma') = 0$. The classic examples of ruled surfaces are: cylinder, cone and hyperbolic paraboloid (doubly ruled).

The helicoid $f(s,t) = (s\cos t, s\sin t, ht)$, $h \in \mathbb{R} \cup \{\pm\infty\}$ is also a ruled surface. Its axis is a vertical geodesic $c(t) := (0,0,ht)$ and it has horizontal geodesics as rulings $\gamma(s) := (s\cos t_0, s\sin t_0, ht_0)$. The pitch is given by the parameter h, it controls the constant rotation-speed. For $h = 0$ it is a horizontal plane and for $h = \pm\infty$ it is a vertical plane. We claim, it is minimal, because the helicoid is invariant under π-rotation about its rulings. Let $\tilde{\gamma}$ be a curve, which is perpendicular to a ruling γ. The normal curvature $\kappa_{\mathrm{norm}}(\tilde{\gamma})$ changes sign under rotation, therefore $\kappa_{\mathrm{norm}}(\tilde{\gamma}) = 0$, i.e. $\tilde{\gamma}$ is a asymptotic direction and perpendicular to γ. With the Euler-curvature-formula ($g(Sv_\alpha, v_\alpha) = \kappa_1\cos^2\alpha + \kappa_2\sin^2\alpha$) we get $g(Sv_{\alpha_i}, v_{\alpha_i}) = 0$, $i = 1,2$ where $\alpha_1 = \alpha_2 - \frac{\pi}{2} \Rightarrow \kappa_1 = -\kappa_2$.

In Riemannian fibrations with geodesic fibres π rotations about horizontal and vertical geodesics are isometries. Therefore, surfaces which are invariant under π-rotations about geodesics are minimal. Hence, we consider surfaces foliated by geodesics.

10.1.1 Vertical planes

A vertical plane is defined as the preimage $\pi^{-1}(c)$ of a geodesic $c \subset \Sigma$. Vertical planes are minimal, since the horizontal lift of c is a geodesic. Therefore, the surface is foliated by geodesics. Moreover, a π-rotation about each geodesic leaves the plane invariant. In a product space we have for example $\{c\} \times \mathbb{R}$ for a geodesic $c \in \Sigma(\kappa)$. In $E(4,1) = \mathbb{S}^3$ a vertical plane is a Clifford torus.

10.1.2 Horizontal umbrellas

Horizontal umbrellas correspond to horizontal planes. They are defined as the exponential map of a horizontal plane in a point $p \in E(\kappa, \tau)$. Therefore, each umbrella consists of all horizontal radial geodesics starting at p. In $\Sigma(\kappa) \times \mathbb{R}$ a horizontal umbrella is totally geodesic, whereas for $t \neq 0$ the surface has non-horizontal tangent spaces except in p. Horizontal umbrellas are minimal. For $\kappa \leqslant 0$ or $\tau = 0$ they are sections. Each surface is of disc type for $\kappa \leqslant 0$ and a sphere for $\kappa > 0$, for example a geodesic 2-sphere in $\mathbb{S}^3$.

10.1.3 Horizontal slices

We interpret a horizontal slice as a horizontal helicoid, where the axis is a horizontal geodesic c and the rulings are the horizontal geodesics, which are perpendicular to c. Horizontal slices are minimal. Topological it is a disc for $\kappa \leqslant 0$, a torus if $\kappa > 0, \tau \neq 0$, and a sphere if $\kappa > 0, \tau = 0$.

10.1.4 Vertical helicoids

Last but not least we consider vertical helicoids $M(s)$. It is family of minimal surfaces, where the axis is a fibre of $\pi : E \to \Sigma$ and the rulings are horizontal geodesics, which rotate along the axis with constant speed s. As in $\mathbb{R}^3$, we have special cases: The surface $M(\tau) \subset E(\kappa, \tau)$ is a vertical plane and $M(\pm\infty)$ are horizontal umbrellas.

10.2 Horizontal helicoid in Nil_3

In the construction of the k-noid with genus 1 from Section 11.2, especially to solve one of the period problems, we use as a barrier the horizontal helicoid $H_\alpha(u,v)$ in Nil, $\left(\mathbb{R}^3, \mathrm{d}x_1^2 + \mathrm{d}x_2^2 + (\mathrm{d}x_3 - x_1\,\mathrm{d}x_2)^2\right)$ by Daniel and Hauswirth [DH09, Section 7]. For $\alpha > 0$ the coordinates of the helicoid H_α are given in terms of the solution ψ of the ordinary differential equation $\psi'^2 = \alpha^2 + \cos^2\psi, \psi(0) = 0$:

$$
\begin{aligned}
x_1 &= \frac{\sinh(\alpha v)}{\alpha(\psi'(u) - \alpha)} \cos\psi(u) \\
x_2 &= -G(u) \\
x_3 &= \frac{-\sinh(\alpha v)}{\alpha(\psi'(u) - \alpha)} \sin\psi(u),
\end{aligned}
$$

where G is defined by $G'(u) = 1/(\psi'(u) - \alpha), G(0) = 0$. In [DH09] was shown that the function ψ is a decreasing odd bijection. There exists a unique $U := U(\alpha) > 0$ with $\psi_\alpha(U) = -\pi/2$, $\psi_\alpha(-U) = \pi/2$. To visualise the surface, we look at three curves in the helicoid:

$$
\begin{aligned}
H_\alpha(-U,v) &= \left(0, G(U), \frac{\sinh(\alpha v)}{\alpha(\alpha - \psi'(-U))}\right) \\
H_\alpha(0,v) &= \left(\frac{\sinh(\alpha v)}{\alpha(\psi'(0) - \alpha)}, 0, 0\right) \\
H_\alpha(U,v) &= \left(0, -G(U), \frac{\sinh(\alpha v)}{\alpha(\psi' - \alpha)}\right).
\end{aligned}
$$

The rulings $H_\alpha(\pm U, v)$ are vertical and define the width $a := G(-U) - G(U)$ of the helicoid. The width is well-defined for the whole helicoid, since $\psi(u + 2U) = \psi(u) - \pi$.

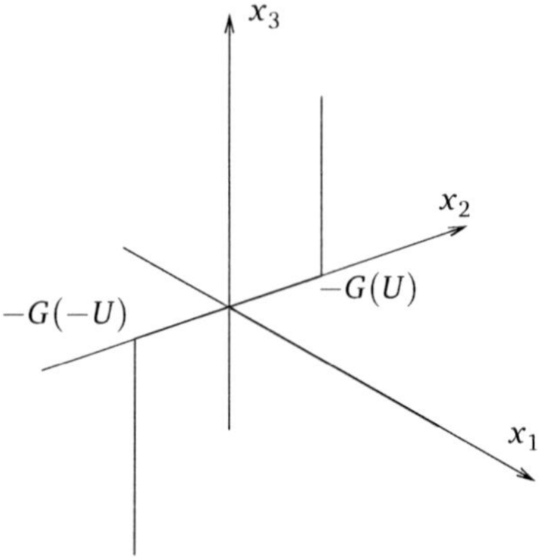

Figure 10.1.: Sketch of a fundamental piece of the horizontal helicoid from Daniel and Hauswirth in Nil, $\upsilon \leqslant 0$.

To ensure that we can consider a helicoid H_a for a given width a, we need the following lemma:

Lemma 10.1. *For $a > 0$, there exists $\alpha > 0$ such that*

$$-2G(U(\alpha)) = a,$$

where $U(\alpha)$ is defined by $\psi_\alpha(U) = -\pi/2$. Furthermore, for $a \to 0$ we have $\alpha \to \infty$.

Proof. The idea of the proof is to show that G is a bijection on $\mathbb{R}$ as a first step. The second step is to show that U is a continuous map to $\mathbb{R}_+$.

Step 1: The function G is odd, so $G(-U) - G(U) = -2G(U)$. For $a > 0$ we show that there exists exactly one $U > 0$ such that $-2G(U) = a$: From $G' = 1/(\psi' - \alpha) < 0$ we know that G is a decreasing function on $\mathbb{R}$. If we assume that G is bounded, i.e. $G(u) \to g \in \mathbb{R}$ for $u \to \infty$, then $G'(u) \to 0$ for $u \to \infty$. But this implies $\psi'(u) - \alpha \to -\infty$ for $u \to \infty$, which is a contradiction because $\psi'^2 = \alpha^2 + \cos^2\psi \leqslant \alpha^2 + 1$ is bounded. Therefore, G is a decreasing bijection on $\mathbb{R}$.

Step 2: We show for $U > 0$ the existence of $\alpha > 0$ such that the solution ψ_α of

$$\psi_\alpha'^2 = \alpha^2 + \cos^2\psi_\alpha, \quad \psi_\alpha(0) = 0$$

satisfies $\psi_\alpha(U) = -\pi/2$.

By applying seperation of variables to the ODE $\psi_\alpha' = -\sqrt{\alpha^2 + \cos^2\psi}$, $\psi(0) = 0$, the solution is then given by the inverse of the elliptic integral of the first kind

$$\int_0^{\psi} -\frac{1}{\sqrt{\alpha^2 + \cos^2\vartheta}}\,\mathrm{d}\vartheta.$$

We are interested in $U(\alpha)$ given by $\psi_\alpha(U) = -\pi/2$:

$$\begin{aligned}
U(\alpha) &= \int_0^{-\pi/2} -\frac{1}{\sqrt{\alpha^2 + \cos^2\vartheta}}\,\mathrm{d}\vartheta \\
&= \int_0^{\frac{\pi}{2}} \frac{1}{\sqrt{\alpha^2 + \cos^2\vartheta}}\,\mathrm{d}\vartheta \\
&= \frac{1}{\sqrt{\alpha^2 + 1}} \int_0^{\frac{\pi}{2}} \frac{1}{\sqrt{1 - \frac{1}{\alpha^2+1}\sin^2\vartheta}}\,\mathrm{d}\vartheta \\
&= \frac{K\left(1/\sqrt{\alpha^2+1}\right)}{\sqrt{\alpha^2+1}},
\end{aligned}$$

where $K(k) = \int_0^{\pi/2} \frac{\mathrm{d}\vartheta}{\sqrt{1-k^2\sin^2\vartheta}} = \int_0^1 \frac{\mathrm{d}t}{(1-t^2)(1-k^2t^2)}$ denotes the complete elliptic integral of the first kind, defined for $k \in [0,1)$, with the special values $K(0) = \pi/2$ and $\lim_{k\to 1} K(k) = \infty$. For $\alpha \to 0$ we have $K\left(1/\sqrt{\alpha^2+1}\right) \to \infty$ and for $\alpha \to \infty$ we have $K\left(1/\sqrt{\alpha^2+1}\right) \to \pi/2$. Therefore, U is continuous because K is. Moreover, for $\alpha \to 0$ we have $U(\alpha) \to \infty$ and for $\alpha \to \infty$ we have $U(\alpha) \to 0$.

Together with the Step 1, this concludes the first part of the lemma. For the second part, notice that $a \to 0$ implies $U \to 0$, since G is a decreasing function with $G(0) = 0$. Furthermore, $U(\alpha) \to 0$ implies $\alpha \to \infty$, since K is bounded. □

We want to express the height b of the conormal η of the helicoid H_a along the vertical rulings depending on the width a and the angle φ in the horizontal plane $\operatorname{span}\{\partial x_1, \partial x_2 + x_1\partial x_3\}$ given by

$$\cos\varphi = \langle \partial x_1, \eta \rangle.$$

The conormal along the horizontal ruling

$$H_\alpha(-U,v) = \left(0, G(U), \frac{\sinh(\alpha v)}{\alpha(\psi'(-U)-\alpha)}\right)$$

is given by

$$\frac{\partial H_\alpha}{\partial u}(-U,v) = \left(\frac{-\sinh(\alpha v)\psi'(-U)}{\alpha(\psi'(-U)-\alpha)}, -G'(-U), 0\right), \quad \nu = \frac{\partial_u H_\alpha}{||\partial_u H_\alpha||}.$$

The conormal along $H_\alpha(-U,v)$ is horizontal, since $x_1 = 0$. We may express φ in terms of $(\alpha, U = U(\alpha))$. By [DH09] we have

$$\psi'(-U) = G'(-U)\cos^2\psi(-U) - \alpha = -\alpha \quad \text{and} \quad G'(-U) = \frac{1}{\psi'(-U)-\alpha} = \frac{-1}{2\alpha}.$$

Therefore

$$\eta = \frac{2\alpha^2}{\sqrt{\alpha^2(\sinh^2(\alpha v)+1)}}\partial_u H_\alpha(-U,v) = \frac{2\alpha}{\cosh(\alpha v)}\partial_u H_\alpha(-U,v)$$

and

$$\cos\varphi = \frac{2\alpha^2\sinh(\alpha v)}{-2\alpha^2\cosh(\alpha v)} = \tanh(-\alpha v),\, v \leqslant 0 \Leftrightarrow v = \frac{-1}{2\alpha}\ln\left(\frac{1+\cos\varphi}{1-\cos\varphi}\right).$$

Using this we get the height b of the conormal η, in terms of φ and $(\alpha, U(\alpha))$ for fixed $\alpha > 0$, as follows:

$$b = \frac{\sinh\left(\frac{-1}{2}\ln\left(\frac{1+\cos\varphi}{1-\cos\varphi}\right)\right)}{2\alpha^2} \leqslant 0.$$

One readily sees that $\varphi \to 0$ implies $b \to -\infty$, and $b \to 0$ when $\varphi \to \pi/2$.

10.2.1 Catenoid in $\mathbb{H}^2 \times \mathbb{R}$

In [DH09] Daniel and Hauswirth also analysed the sister surface C_α of the horizontal helicoid H_α. We call it a *horizontal catenoid.* It is a MC 1/2 annulus in $\mathbb{H}^2 \times \mathbb{R}$. They prove that it is a properly embedded, and its level zero curves have two connected components, each behaving like a Bryant surface generatrix.

10.3 Constant mean curvature k-noids in $\Sigma(\kappa) \times \mathbb{R}$

In [GK10, Section 5.] Große-Brauckmann and Kusner constructed a one-parameter family of surfaces with constant mean curvature $H \geqslant 0$ in $\Sigma(\kappa) \times \mathbb{R}$, which have k ends and dihedral symmetry. They solved a Plateau problem of disc type in $E(\kappa + 4H^2, H)$, its sister generates the CMC surface by reflections about horizontal and vertical planes. We use the minimal disc $M = M(a,k)$ as a barrier in our constructions.

The minimal surface M is the limit of a sequence of compact Plateau solutions $M_{(r,s)}$, which represent sections in $E(\kappa + 4H^2, H)$. Each minimal disc $M_{(r,s)}$ is bounded by horizontal and vertical geodesics, see Figure 10.1. Let $\Gamma_{(r,s)}$ denote the boundary. The minimal surface $M_{(r,s)}$ is a section of the trivial line bundle $\pi \colon \Omega_r \subset E(\kappa + 4H^2, H) \to \Delta_r$, where $\Omega_r := \pi^{-1}(\Delta_r)$ is a mean convex domain, which is defined as the preimage of a triangle $\Delta_r \subset \Sigma(\kappa + 4H^2)$. The triangle Δ_r is given by a hinge of lengths a and r, enclosing an angle π/k. The parameter a determines the length of the horizontal edge in the boundary of M, it corresponds to the necksize in the CMC sister.

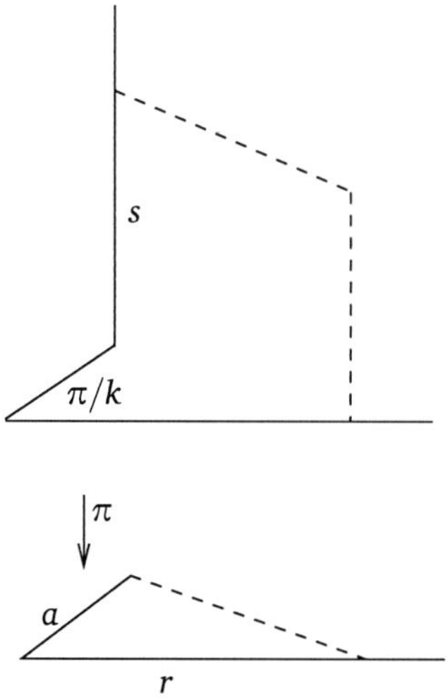

Figure 10.1.: The boundary of the minimal disc in $E(\kappa + 4H^2, H)$.

The Jordan curves $\Gamma_{(r,s)}$ converge to the boundary Γ of the desired minimal disc M. Große-Brauckmann and Kusner proved the following theorem:

Theorem 10.2 ([GK10]). *Suppose $\kappa + 4H^2 \leqslant 0$. Then there exist minimal surfaces $M = M(a,k)$ with boundary Γ, having the following properties:*

- *M is a section projecting to $\Delta := \lim_{r\to\infty} \Delta_r$,*
- *M extends without branch points by the Schwarz reflection about the edges of Γ.*

11 New surfaces

11.1 Constant mean curvature $2k$-noids

We construct a 2-parameter family of surfaces with CMC $H \in [0, 1/2]$ in $\Sigma(\kappa) \times \mathbb{R}$ with $2k$ ends and dihedral symmetry. Each surface has k vertical symmetry planes and one horizontal one, where $k \geqslant 2$. The idea is to solve an improper Plateau problem of disc type in $E(\kappa + 4H^2, H)$, where $\kappa + 4H^2 \leqslant 0$, and the disc is bounded by geodesics. Its sister disc in $\Sigma(\kappa) \times \mathbb{R}$ generates an MC H surface by reflections about horizontal and vertical planes.

For $\kappa = -1$ and $H = 1/2$, it corresponds to a minimal surface in $\mathrm{Nil}_3(\mathbb{R}) = E(0, 1/2)$; if $0 < H < 1/2$ the MCH surface results from a minimal surface in $\widetilde{\mathrm{SL}}_2(\mathbb{R})$ and finally for $H = 0$ the surfaces are conjugate minimal surfaces in $\Sigma(\kappa) \times \mathbb{R}$, $\kappa \leqslant 0$.

11.1.1 Boundary construction

In $\Sigma(\kappa) \times \mathbb{R}$ the desired boundary is not connected. It consists of two components: The first component is a vertical mirror curve consisting of two rays, each lying in a vertical plane. The two planes form an angle $\varphi = \pi/k$, $k \geqslant 2$. The second component is a mirror curve in a horizontal plane. In Section 7.2 we discussed the relation between vertical and horizontal mirror curves and their geodesic sisters. By Proposition 7.7 the sister surface in $E(\kappa + 4H^2, H)$ is bounded by a geodesic contour $\Gamma := \Gamma_{d,\alpha}$: The horizontal mirror curve corresponds to a vertical geodesic and the vertical mirror curves are related to horizontal geodesics enclosing an angle π/k. The relative position of the vertical and horizontal components determines the $2k$-noid.

The distance d of the vertical geodesic to the vertex of the two horizontal geodesics is well-defined and realised by the length of a horizontal geodesic γ. Its length is equal to the length of its projection $\pi(\gamma)$ to $\Sigma(\kappa + 4H^2)$, since it is a horizontal geodesic and the projection π is a Riemannian fibration. The same holds for the angle α enclosed by γ and one of the horizontal rays. Since the sister surfaces are isometric, it is consistent to call the 2-parameter family of CMC surfaces which we will obtain $\widetilde{M}_{d,\alpha}$.

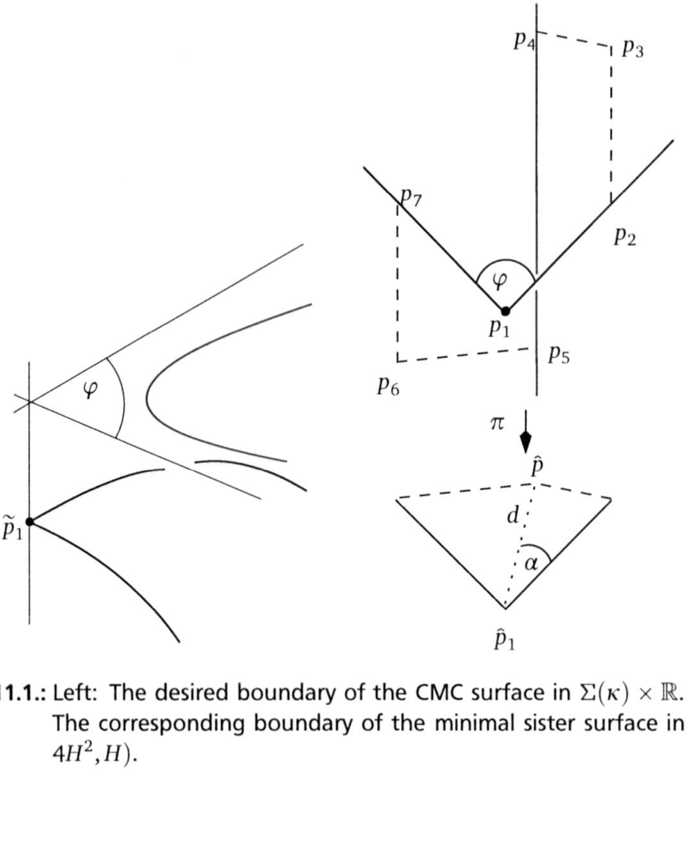

Figure 11.1.: Left: The desired boundary of the CMC surface in $\Sigma(\kappa) \times \mathbb{R}$. Right: The corresponding boundary of the minimal sister surface in $E(\kappa + 4H^2, H)$.

To construct a minimal surface that is bounded by Γ, we truncate the infinite contour Γ and get Jordan curves Γ_n, $n > 0$. We consider a geodesic quadrilateral $\Delta_n := \Delta_n(d, \alpha)$ in $\Sigma(\kappa + 4H^2)$: Two edges of length n form an angle π/k and intersect in point $\hat{p}_1$. Furthermore, its diagonal in $\hat{p}_1$ has length d and encloses an angle $\alpha \leqslant \varphi/2$ to one side. Let $\hat{p}$ denote the endpoint of the diagonal. We consider the horizontal lift of Δ_n starting in $\hat{p}$ and going in positive direction. We label the endpoints with $\widetilde{\Delta}_n(0) = p_5$ and $\widetilde{\Delta}_n(l) = p_4$, by Lemma 1.4 the signed vertical distance is $d(p_5, p_4) = 2H\,\text{area}(\Delta_n)$ and therefore, in positive ξ-direction. Now we translate the horizontal edge that ends in p_4 in positive ξ-direction by n and call the endpoint p_3. Furthermore, we translate the horizontal edge that starts in p_5 in $-\xi$-direction by n and call the endpoint p_6. After the vertical translation we have

$$d(p_5, p_4) = 2H\,\text{area}(\Delta_n) + 2n.$$

Hence, p_5p_4 is in positive ξ-direction. We complete the Jordan curve Γ_n by adding two vertical edges of length n in p_3 and p_6 and label the intersections with the horizontal edges p_2 and p_7 respectively. See Figure 11.1.

The polygon Γ_n has six right angles and one angle π/k; the quadrilateral Δ_n is not necessarily convex for large n.

For $n \to \infty$ we have $\Gamma_n \to \Gamma$; this contour can be constructed by the union of quadrilaterals

$$\Delta := \Delta(d, \alpha) = \bigcup_{n>0} \Delta_n.$$

Note that Δ is non-convex.

11.1.2 Plateau solutions

The idea is to consider the Plateau solutions for Γ_n and to take their limit $n \to \infty$. We show that there exists a domain Ω_n with mean convex boundary, such that $\Gamma_n \subset \partial\Omega_n$. By [HS88] the solution of the Plateau problem for Γ_n is a minimal disc. Then it is shown that the sequence is controlled by a curvature estimate.

Proposition 11.1. *The special Jordan curve $\Gamma_n \subset E(\kappa + 4H^2, H)$ bounds a Plateau solution $M_n \subset E(\kappa + 4H^2, H)$ for large $n \in \mathbb{N}$, which extends without branch points by Schwarz reflection about the edges of Γ_n.*

Proof. We define the domain Ω_n with mean convex boundary as the intersection of five domains: two of them have horizontal umbrellas as boundaries, two have vertical planes as boundaries and the last domain has four fundamental pieces of the CMC n-noids from [GK10] (GK-surfaces) in its boundary:

1. Take the halfspaces above the horizontal umbrella U_5 in p_5 and below the horizontal umbrella U_4 in p_4, see Subsection 10.1.2 for the definition. Below resp. above means in negative resp. positive ξ-direction. We call the intersection of the two halfspaces a *horizontal slab*. The umbrellas are with respect to the same fibre and therefore are parallel sections with vertical distance $d(U_5, U_4) = d(p_5, p_4)$. If $U_{4/5} \cap \Gamma_n \neq \emptyset$ we redefine Γ_n by translating $p_{4/5}$ in $\pm\xi$ direction by factor $c_{4/5}$. Since $U_{4/5}$ are sections, they are graphs above/below the horizontal geodesics p_1p_2 and p_1p_7. Therefore we find constants $c_{4/5} > 0$ such that the new boundary curve, we call it again Γ_n does not intersect the umbrellas. The horizontal slab is a barrier for Γ_n, for all $n \in \mathbb{N}$.

2. Furthermore, consider the vertical halfspaces defined by the horizontal arcs p_1p_2 and p_1p_7, such that Γ_n lies inside.

3. The last domain is based on GK-surfaces: The idea is to consider a mean convex set sandwiched between two copies of a symmetric CMC $2k$-noid piece. We claim that we can orient them and choose their parameters, the necksize a and the number of ends, such that they are barriers for Γ_n. Their position relative to Γ_n is given by a rotation angle δ and the vertical distances $h_\pm$ from the horizontal umbrella U in p_1. We call the vertical distance to U *height*.

   We take four fundamental patches $M_{d,2k}$ from [GK10, Theorem 5.2] and orient them, such that their axes coincide with p_4p_5. We require their horizontal boundaries to lie in the horizontal umbrella U, see Subsection 10.1.2. By rotation about p_4p_5 by an small angle $\pm\delta$ followed by vertical translations by $h_\pm$ we construct a mean convex domain S. Each GK-surface $S_\pm$ consists of two fundamental domains generated by Schwarz reflection about the bounded horizontal edge.

   We define S_+ first: We choose an orientation such that the projection of its horizontal edge of length d coincides with the diagonal of the quadrilateral in the projection. Afterwards we translate the surface in ξ-direction by h_+. There exists $N \in \mathbb{N}$ such that for all $n \geqslant N$ the surface S_+ does not intersect Γ_n for all $h_+ > 0$, because S_+ is graph above the projection of the horizontal edges of Γ_n. In the projection the horizontal hinge of Γ_n encloses an angle $\delta := \varphi/2 - \alpha > 0$ with the horizontal hinge of ∂S_+.

   The other GK-patch is rotated about p_4p_5 such that in the projection the two edges of length d enclose an angle δ. Afterwards we translate the surface by h_- in $-\xi$-direction. The surface S_- is a graph below a bounded component

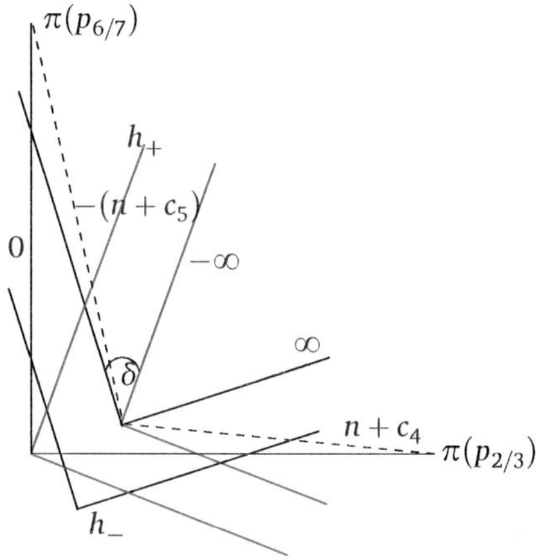

Figure 11.2.: The projection of the three boundaries of M_n, S_+ (green) and S_+ (blue) to $\Sigma(\kappa + 4H^2)$. The heights of the horizontal geodesics are labelled in red. Here $k = 2$ and $\kappa + 4H^2 = 0$.

of p_1p_2, therefore exists $h_- > 0$ such that $S_- \cap \overline{p_1p_2} = \varnothing$. Furthermore, it is a graph below p_1p_7, where the surface lies below the horizontal umbrella at height h_-. The other edges of Γ_n are uncritical for n large enough.

It remains to show that for every $p \in p_4p_5$ the opening angle ψ of the tangent cone T_pC of S is less than π. This is clear for p at height $|h| > \max\{h_\pm\}$. The angle has its maximum $\psi(h_+,h_-)$ in height $(h_+ - h_-)/2$, it depends on h_+ and h_- and is bounded by $\psi_{\sup} := 2(\pi - \varphi - \epsilon) + \delta$ where $\epsilon \geqslant 0$ denotes the defect depending on $\Sigma(\kappa + 4H^2)$. We consider the level curves of S_+ and S_- in height h. The level curves define angles $\beta_+(h)$ and $\beta_-(h)$ given in the projection by the angle of the projected conormal in height h of $S_\pm$ and the edge of length d of the corresponding surface. We know $0 \leqslant \beta_\pm < \pi - \pi/(2k) - \epsilon$. Therefore, we have

$$\psi(h_+,h_-) = \beta_+\left(\frac{h_+ - h_-}{2}\right) + \beta_-\left(\frac{h_+ - h_-}{2}\right) + \delta.$$

But h_+ was chosen independently of n, δ and h_-, moreover for $h_+ \to 0$ we have $\beta_+\big((h_+ - h_-)/2\big) \to 0$. Hence we conclude

$$\psi(h_+,h_-) \to \beta_-\left(\frac{h_+ - h_-}{2}\right) + \delta < \pi - \epsilon \leqslant \pi.$$

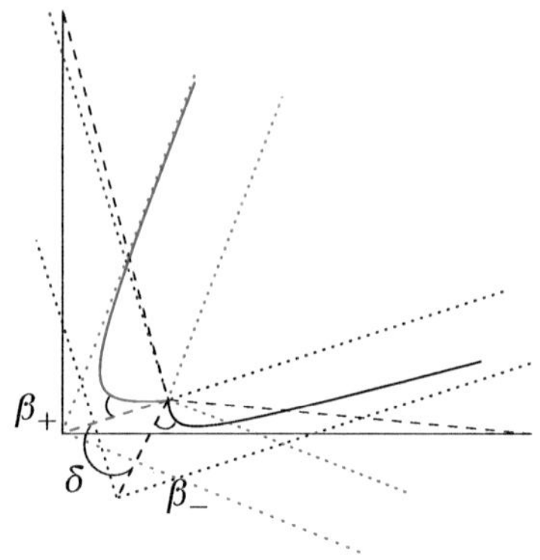

Figure 11.3.: Level curves: The solid lines sketch the intersection of the horizontal umbrella U with M_n and the GK-surfaces (green and blue). The dashed lines indicate the remaining boundaries. Here $k = 2$.

We summarise: Γ_n lies in-between two copies of a CMC $2k$-noid, i.e. there exists $N \in \mathbb{N}$ such that $(S_+ \cup S_-) \cap \Gamma_n = p_4 p_5$ for $n \geqslant N$.

We complete the last barrier by subsets of the two horizontal umbrellas at heights $h_\pm$ given by the edges of the GK-patches; it defines the boundary of the halfspace S.

We define the mean convex domain Ω_n as the intersection of the five halfspaces.

Since Γ_n lies in the boundary of a mean convex domain, the existence of an embedded minimal surface M_n of disc-type with boundary Γ_n follows from Hass and Scott (see Theorem 8.2).

To see that M_n extends by Schwarz reflection about all edges of Γ_n, we consider the finite group $G_p \subset \mathrm{Iso}(E)$ generated by π-rotations about all edges of Γ_n, which contain p. We claim that the images of Ω_n are disjoint under the group G_p, at least in a neighbourhood of p. For $p \notin \overline{p_4 p_5}$ the intersection $\bigcap_{g \in G_p} g(\Omega_n) = \emptyset$ and for $p \in \overline{p_4 p_5}$ we have seen before that for the not-translated GK-surfaces the horizontal angle spanned by $\partial\Omega_n$ is less than $\pi/2$ for all p. By construction, the apex angle of the tangent cone is still less than π after the translations. Therefore, the images of the neighbourhood of p under the group G_p are disjoint. By Proposition 8.4 M_n extends by Schwarz reflection. □

Remark 10. The definition of Ω_n would be more direct if we could define it as the intersection of halfspaces. A vertical plane/horizontal umbrella separates $E(\kappa+4H^2,H)$ into two connected components, but two fundamental patches of a GK-surface do not separate $E(\kappa+4H^2,H)$ in two connected components. This is because of the normal turning along the vertical geodesic. To get several connected components, we have to use $S_+ \cup S_-$, but their boundary would not be smooth anymore.

Lemma 11.2. *The Plateau solution M_n is a section over a simply connected domain Δ_n enclosed by $\partial\Delta_n := \pi(\Gamma_n)$ and unique among all Plateau solutions with the prescribed boundary values for each $n \in \mathbb{N}$.*

Proof. We show that M_n does not have any vertical tangent planes. Then Lemma 8.5 implies that M_n is a section.

Proof by contradiction: Suppose there exists a vertical plane V that is tangent to M_n at some $p \in M_n$. We consider the intersection $V \cap \overline{M_n}$: Since M_n and V are both minimal but not identical, their intersection $M_n \cap V$ is a union of analytic curves ending on $\partial M_n = \Gamma_n$ (See the proof of Proposition 9.3 for details.). At p at least two of them meet. Assume two curves crossing at p extend to a loop $\gamma \subset M_n \cap V$, then the precompact component of $M_n \backslash \gamma$ would coincide with V. By the maximum principle then $M_n \equiv V$, which is impossible since $\Gamma_n \not\subseteq V$. Therefore, $M_n \cap V$ cannot contain a loop and the analytic curves have at least four endpoints on Γ_n.

Let us consider the intersection of the vertical plane with the mean convex domain $\Omega_n \cap V$. It might consist of more than one connected component. For the connected component V_p containing p, we know that $\Gamma_n \cap V_p$ has at most two connected components. So at least two endpoints of $M_n \cap V$ form a loop in $V \cup \overline{M_n}$, which by assumption is excluded. Therefore, we get a contradiction, i.e. there exists no vertical tangent plane.

Since the minimal surface M_n is a section $s\colon \Delta_n \to E$, it is the solution of an elliptic partial differential equation. The uniqueness follows from the maximum principle for elliptic partial differential equations. See Proposition 8.7. □

Now we can take the limit $n \to \infty$:

Theorem 11.3. *There exists a minimal surface $M_\infty \subset E(\kappa+4H^2,H)$, $\kappa+4H^2 \leqslant 0$ which is a section over Δ and extends without branch points by Schwarz reflection across its edges.*

Proof. The sequence of surfaces M_n satisfies a uniform local area bound, since each M_n is a section. Moreover the closure of each mean convex domain $\overline{\Omega}_n$ is compact, their sequence is nested $\Omega_{n-1} \subset \Omega_n$ and $\Omega := \cup_{n>0}\overline{\Omega}_n$ is a non-compact closed manifold with boundary. The limit $\Gamma := \lim_{n\to\infty}\Gamma_n$ is a non-closed piecewise C^1-curve in Ω. The sequence $\Gamma_n \subset \overline{\Omega}_n$ of closed Jordan curves satisfies $\Gamma_k \cap \Omega_n = \Gamma \cap \Omega_n$ if $k > n$. Therefore, Theorem 9.2 applies and there exists a limit, a minimal surface $M_\infty \subset \Omega$ of disc-type with $\partial M_\infty = \Gamma$.

Since the surfaces M_n represent sections the limit surface M_∞ is a vertical plane or a section over $\Delta = \cup_{n>0}\Delta_n = \cup_{n>0}\pi(M_n)$ by Proposition 9.3. But since for every $p \in \Delta$ the sequence $p_n := \pi^{-1}(p) \in M_n$ is bounded by Ω, it cannot be a vertical plane. Finally by Proposition 8.4 M_∞ extends without branch points by Schwarz reflection about the edges of Γ. □

The minimal surface M_∞ is a fundamental piece of a $2k$-noid; we reflect its sister surface to construct a CMC surface in $\Sigma(\kappa) \times \mathbb{R}$:

Theorem 11.4. *For $H \in [0, 1/2]$ and $k \geqslant 2$ there exists a two-parameter family $\left\{\widetilde{M}_{d,\alpha} : d > 0, 0 < \alpha \leqslant \pi/(2k)\right\}$ of constant mean curvature H surfaces in $\Sigma(\kappa) \times \mathbb{R}$, $\kappa \leqslant 0$, such that:*
- *$\widetilde{M}_{d,\alpha}$ has k vertical mirror planes enclosing an π/k-angle,*
- *$\widetilde{M}_{d,\alpha}$ has one horizontal mirror plane and*
- *for $\alpha = \pi/(2k)$ the surface $\widetilde{M}_{d,\alpha}$ is a GK-surface with $n = 2k$.*

Proof. By [Dan07] (see Theorem 7.1) the fundamental piece M_∞ has a sister surface $\widetilde{M}_\infty$ with constant mean curvature H in $\Sigma(\kappa) \times \mathbb{R}$. By construction, $\widetilde{M}_\infty$ has one horizontal and two vertical curves in mirror planes; the two vertical mirror planes enclose an angle π/k. Schwarz reflection about those planes extends the surface to a complete MC H surface $\widetilde{M}_{d,\alpha}$ with $2k$ ends. The MC H-surface $\widetilde{M}_{d,\alpha}$ consists of $4k$ fundamental pieces $\widetilde{M}_\infty$. □

In $\mathbb{H}^2 \times \mathbb{R}$ we can control the horizontal mirror curves and the geometry of the ends. Therefore, we can prove Alexandrov-embeddedness:

Theorem 11.5. *For $H = 1/2$, $\kappa = -1$ the two-parameter family $\{\widetilde{M}_{d,\alpha} : d > 0, 0 < \alpha \leqslant \pi/(2k)\}$ of constant mean curvature $1/2$ surfaces in $\mathbb{H}^2 \times \mathbb{R}$ is Alexandrov-embedded for $k \geqslant 2$.*

Proof. We claim that $\widetilde{M}_{d,\alpha}$ is Alexandrov-embedded. We show the fundamental piece $\widetilde{M}_\infty$ is embedded away from the ends and stays in the subset of $\mathbb{H}^2 \times \mathbb{R}$ that is bounded by the mirror planes. We discuss each boundary arc, since $\widetilde{M}_\infty$ is transverse to the fibres, the Alexandrov-embeddedness of the fundamental domain follows. Therefore, the complete surface $\widetilde{M}_{d,\alpha}$ is Alexandrov-embedded.

Let c_1 denote one of the horizontal arcs of ∂M_∞ and $\widetilde{c}_1$ its sister curve in $\partial\widetilde{M}_\infty$. The vertical plane given by c_1 is a barrier for M_∞. Therefore, $\langle \nu, \xi \rangle \in (0, 1]$ if we choose the up-pointing normal ν. If we require (c', η, ν) to be positively oriented, then c_1' is well-defined. The normal is nowhere horizontal by the Hopf boundary point lemma. By Daniel's correspondence Theorem (7.1) we have the same relation for the sister surface $\widetilde{M}_\infty$

$$\langle \widetilde{\nu}, \widetilde{\xi} \rangle \in (0, 1].$$

The sister curve $\widetilde{c}_1$ is a mirror curve in a vertical plane. The curve is graph and therefore embedded. The same holds for the other horizontal curve and its vertical sister curve.

Along one of the horizontal geodesics we have $\langle \eta, \xi \rangle \geqslant 0$ and along the other we have $\langle \eta, \xi \rangle \leqslant 0$. By the first order description of the sister surfaces we know that the projection of the vertical vector field ξ on the tangent plane rotates by $\pi/2$ under conjugation. If c_1 is the horizontal geodesic with $\langle \eta, \xi \rangle \leqslant 0$, then we have $\langle \widetilde{c}_1', \widetilde{\xi} \rangle \geqslant 0$. By construction, we know that $\langle \eta, \xi \rangle \to -1$ in the end along c_1, i.e. for $t \to -\infty$. Therefore, in the sister surface for the corresponding mirror curve we have $\langle \widetilde{c}_1', \widetilde{\xi} \rangle \to 1$ for $t \to -\infty$. In other words, the mirror curves come from $-\infty$ and go to $-\infty$.

With the arguments of Hauswirth, Rosenberg and Spruck in the proof of Theorem 1.2 in [HRS08] we know that each divergent sequence (p_n) in $\widetilde{M}$ with $\langle \widetilde{\nu}(p_n), \widetilde{\xi} \rangle \to 0$ has a limit in the boundary in the projection: $\pi(p_n) \to \partial\mathbb{H}^2$. The idea of the proof is to show that if the sequence has a horizontal normal in the limit and a limiting point in the projection, then the whole surface is converging to a horocylinder and stays on one side. Hence, by the half-space theorem it is a horocylinder which is a contradiction.

Moreover, we have to check if the surface stays in a vertical halfspace of $\mathbb{H}^2 \times \mathbb{R}$. We define the horizontal mirror plane of $\widetilde{M}_\infty$ to be $\mathbb{H}^2 \times \{0\}$. Since $\langle \widetilde{\nu}, \widetilde{\xi} \rangle = 1$ in the intersection of the vertical mirror planes, it follows from the maximum principle that $\widetilde{M}_\infty$ has no maximum above $\mathbb{H}^2 \times \{0\}$.

Assume there is a curve $\widetilde{c}$ in $\widetilde{M}$ whose third component goes to infinity with $\langle \widetilde{c}', \widetilde{\xi} \rangle \to 1$ for $t \to \infty$. The first order description implies $\langle \eta, \xi \rangle \to -1$ as $t \to \infty$ along the sister curve in the minimal surface. But we have chosen the up-pointing normal and (c', η, ν) is positively oriented, which is a contradiction.

Let c_2 denote the vertical geodesic parametrized in $-\xi$-direction. The surface M_∞ is contained in a mean convex domain Ω, whose opening angle along c_2 in the projection is $2(\pi + \delta) - \varphi = 2(\pi - \alpha)$, where $0 < \alpha \leqslant \varphi/2$ depends on the chosen parameters. The normal ν along c_2 is horizontal and rotates monotonically, because M_∞ is a section. There exists $p := c_2(0) \in c_2$, such that

$$\operatorname{twist}_{+\infty} := \lim_{s\to+\infty} \int_0^s t(r) + 1/2\,\mathrm{d}r < \pi \quad \text{and}$$

$$\operatorname{twist}_{-\infty} := \lim_{s\to-\infty} \int_s^0 t(r) + 1/2\,\mathrm{d}r < \pi,$$

since twist $=$ twist$_{+\infty}$ $+$ twist$_{-\infty}$ is less than the opening angle of the mean convex domain in the projection. We consider the horocycle fibration of $\mathbb{H}^2$ given by $\widetilde{\nu}(0)$ as in Proposition 7.11. By Proposition 7.11 the angle ϑ the sister curve encloses with the fibration in $\mathbb{H}^2$ is smaller than twist$_{\pm\infty}$. Hence, each half of $\widetilde{c}_2$ ($\widetilde{c}_2(t)$ for $t \geqslant 0$ and for $t \leqslant 0$) is embedded. Moreover, for each half we wrap the surface between two helicoids H_b and H_a with horizontal axes defined by Daniel and Hauswirth in [DH09], which are barriers from below and above in the sense of a Riemannian fibration. See Section 10.2 for the description of the helicoids. We may orient them such that their rulings in ξ-direction coincide and translate them in fibre-direction such that the three surfaces intersect in the vertical geodesic only. Since they are barriers their torsions along each component c_2 are related:

$$t_b \leqslant t \leqslant t_a.$$

Daniel and Hauswirth showed that the sister surface of the horizontal helicoid is a 1/2 catenoid in $\mathbb{H}^2 \times \mathbb{R}$. Moreover the sister curve of the vertical geodesic is a horizontal mirror curve that behaves like a generatrix of a Bryant surface in hyperbolic 3-space, i.e. $\widetilde{k}_{a/b} \to 1$. By Section 7.2 we know $\widetilde{k} = -t + H$, therefore

$$\widetilde{k}_b \geqslant \widetilde{k} \geqslant \widetilde{k}_a$$

and finally $\widetilde{k} \to 1$.

To finish the proof we have to show that the surface lies in-between the two vertical symmetry planes: We have seen that the projection of the vertical symmetry curves is converging to $p_{1,2} \in \partial\mathbb{H}^2$. The surface is at least locally to the side of the normal, since $\widetilde{k} < 1$ along $\widetilde{c}_2$ by the graph property of the minimal surface along c_2 . Therefore, since it is also a graph, it is defined on a domain Ω given by $\pi(\widetilde{c}_1) \cup \widetilde{c}_2$. □

11.2 Constant mean curvature k-noids with genus 1

We construct surfaces with MC $1/2$ in $\mathbb{H}^2 \times \mathbb{R}$ with k ends and genus 1. Each surface has k vertical symmetry planes and one horizontal symmetry plane, where $k \geqslant 3$. The idea is to solve a Plateau problem of disc type in $\mathrm{Nil}_3(\mathbb{R})$, where the disc is bounded by geodesics. Its sister disc in $\mathbb{H}^2 \times \mathbb{R}$ generates an MC $1/2$ surface by reflections about horizontal and vertical planes. The problem is to define the geodesic contour such that the sister has the desired properties.

11.2.1 Boundary construction

In $\mathbb{H}^2 \times \mathbb{R}$ the desired boundary is connected and consists of four mirror curves in three symmetry planes; the two vertical symmetry planes form an angle π/k, see Figure 11.1.

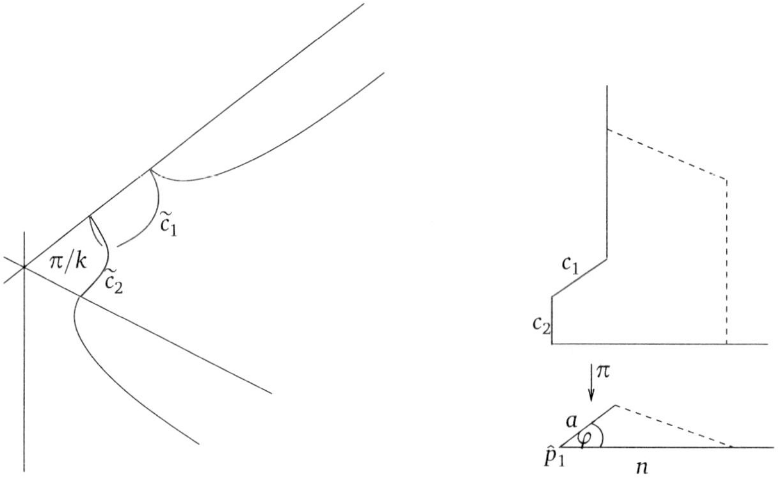

Figure 11.1.: The desired boundary of the $1/2$-surface in $\mathbb{H}^2 \times \mathbb{R}$ and its minimal sister surface in Nil.

By [GK10] the sister surface is bounded by a geodesic contour $\Gamma := \Gamma(a, b, \varphi)$: The horizontal mirror curves correspond to vertical geodesics and the vertical mirror curves correspond to horizontal geodesics, their projections enclose an angle $\varphi \in (0, \pi)$.

The length $a > 0$ of the finite horizontal geodesic c_1 determines the asymptotic parameter of the ends of the k-noid, called the necksize. The length $b > 0$ of the vertical finite edge c_2 defines the size of the hole of the k-noid: For $b \to 0$ the k-noid is close to the non-degenerate k-noid of [GK10]. The angle φ measures the curvature of $\tilde{c}_2$. The parameters will be determined by solving the period problems.

To construct a minimal surface that is bounded by Γ, we truncate the infinite contour Γ and get Jordan curves Γ_n, $n > 0$. We consider a geodesic triangle Δ_n in the base manifold $\mathbb{R}^2$ of the Riemannian fibration of $\mathrm{Nil}_3(\mathbb{R})$: Two edges of lengths a and n form an angle φ and intersect in a point $\hat{p}_1$. We lift $\hat{p}_1$ and the corresponding edge of length n horizontally and label the vertices with p_1 and p_6. Then we add a vertical arc of length b at p_1 in fibre-direction ξ and label its endpoint with p_2. We lift the edge of length a of the base triangle in $\mathbb{R}^2$ horizontally, such that it ends in p_2; the other vertex is labelled by p_3. We add another vertical edge in fibre-direction ξ: it starts in p_3, has length n^2 and its end vertex is called p_4. By lifting the remaining edge of the base triangle horizontally and inserting another vertical edge with endpoints p_5 and p_6 we complete the special Jordan curve Γ_n.

In general the vertical distances do not sum up: $d(p_1, p_2) + d(p_3, p_4) \neq d(p_5, p_6)$, but we claim p_6p_5 is in fibre direction. We consider an arc length parametrization γ of $\partial\Delta_n \subset \mathbb{R}^2$ which runs counter-clockwise. By Lemma 1.4 the vertical distance of its horizontal lift is $\mathrm{area}(\Delta_n)$. By construction we get

$$d(p_6, p_5) = b + n^2 - \mathrm{area}(\Delta_n).$$

Since $\mathrm{area}(\Delta_n)$ grows linearly, there exists $N \in \mathbb{N}$ such that for all $n \geqslant N$: $d(p_6, p_5) > 0$ and $d(p_6, p_5) \to \infty$ for $n \to \infty$.

The polygon Γ_n has six right angles; its projection Δ_n is convex for every n and has one fixed angle $\varphi \leqslant \pi$ independent of n. We define a mean convex set $\Omega_n := \pi^{-1}(\Delta_n) \subset \mathrm{Nil}_3(\mathbb{R})$.

For $n \to \infty$ we have $\Gamma_n \to \Gamma$. To control the Plateau solution for Γ, we solve the Plateau problem for Γ_n and consider their limit.

Lemma 11.6. *The special Jordan curve $\Gamma_n \subset \mathrm{Nil}_3(\mathbb{R})$ bounds a unique Plateau solution $M_n \subset \Gamma_n$. It is a section over Δ_n and extends without branch points by Schwarz reflection about the edges of Γ_n.*

Proof. Since Δ_n is convex and $\partial\Delta_n$ consists of geodesics, its preimage $\Omega_n = \pi^{-1}(\Delta_n)$ is a mean convex set and by construction we have $\Gamma_n \subset \partial\Omega_n$. Therefore, the Plateau problem is solvable by Theorem 8.2. Moreover, by Chapter 8 the solution M_n is a unique section of the trivial line bundle $\pi\colon \Omega_n \to \Delta_n$ that extends as an immersion across Γ_n. □

11.2.2 Plateau solutions

The Plateau solutions M_n are contained in a nested sequence of mean convex sets, hence we are able to solve the Plateau problem for Γ:

Theorem 11.7. *There exists a unique minimal surface $M(a, b, \varphi) \subset \mathrm{Nil}_3(\mathbb{R})$, which is a section over $\Delta := \pi(\lim \Gamma_n)$ and extends without branch points by Schwarz reflection about its edges for all $a, b > 0$ and $\varphi \in (0, \pi)$.*

Proof. Two copies of the minimal k-noid from [GK10] bound the sequence M_n from below and above if they coincide along the vertical axis and are obtained by a vertical translation. Therefore, we can apply Theorem 9.2 to the Plateau solutions M_n. Thus, there exists a minimal surface $M(a, b, \varphi)$ of disc-type for all $a, b > 0$ and $\varphi \in (0, \pi)$, which is regular in the interior and $\partial M = \Gamma$. Since Γ consists of geodesics and has bounded geometry, Schwarz reflection applies and M is smooth at the boundary.

By Proposition 9.3 the limit is a section, since $\lim \Gamma_n$ is not contained in a vertical plane. □

11.2.3 Period problems

To construct an MC 1/2 surface with genus 1 and certain symmetries we have to solve two period problems. One period is given by the vertical distance of the two horizontal mirror curves. The second period is angular and determined by the geometry of the finite horizontal curve.

To solve the first period problem for the MC 1/2 surface $\widetilde{M} \subset \mathbb{H}^2 \times \mathbb{R}$, i.e. to construct a minimal surface $M \subset \mathrm{Nil}_3(\mathbb{R})$ such that the two horizontal components of its sister surface lie in the same horizontal plane, we consider the mirror curve in the vertical plane with finite length a in $\partial\widetilde{M}$, and call it $\widetilde{c}_1$. The period is given by $p = \int \langle \widetilde{c}_1', \widetilde{\xi} \rangle_{\mathbb{H}^2 \times \mathbb{R}}$, where $\widetilde{\xi}$ is the vertical vector field of $\mathbb{H}^2 \times \mathbb{R}$. Since we have a first order description for the vertical parts of vector fields, we consider the vector field $\widetilde{T}$ on $\widetilde{M}$ given by $\widetilde{\xi} - \langle \widetilde{\xi}, \widetilde{\nu} \rangle \widetilde{\nu}$, where $\widetilde{\nu}$ denotes the normal of $\widetilde{M}$. It is the tangential projection of the vertical vector field and rotates by $\pi/2$ in the tangent plane under conjugation:

$$\mathrm{d}f^{-1}(T) = J\,\mathrm{d}\widetilde{f}^{-1}(\widetilde{T}),$$

where f and $\tilde{f}$ denote the corresponding parametrisations of M and $\tilde{M}$. Therefore, we have an analogous formulation of the period on M:

$$\begin{aligned} p(\tilde{M}) &= \int_{\tilde{c}_1} \langle \tilde{c}_1', \tilde{\xi} \rangle_{\mathbb{H}^2\times\mathbb{R}} = \int \langle \mathrm{d}\tilde{f}(\gamma'), \tilde{T} \rangle_{\mathbb{H}^2\times\mathbb{R}} \\ &= \int \langle \mathrm{d}\tilde{f}(\gamma'), \mathrm{d}\tilde{f}(J^{-1}\,\mathrm{d}f^{-1}(T)) \rangle_{\mathbb{H}^2\times\mathbb{R}} = \int \langle J\gamma', \mathrm{d}f^{-1}(T) \rangle_{\mathrm{Nil}_3(\mathbb{R})} \\ &= \int_{c_1} \langle \eta, \xi \rangle_{\mathrm{Nil}_3(\mathbb{R})} = p(M). \end{aligned}$$

As seen above, we have a Plateau solution $M(a,b,\varphi)$ for all $a,b>0$ and $\varphi \in (0,\pi)$. The first period problem is solvable for each $a>0$ and $\varphi \in (0,\pi/2)$:

Proposition 11.8. *For each $a>0$ and $\varphi \in (0,\pi/2)$ there exists $b(a,\varphi)>0$ such that $p(M(a,b(a,\varphi),\varphi)) = 0$.*

Proof. Let $a>0$ and $\varphi \in (0,\pi/2)$ be fixed. By Theorem 11.7 we have a unique Plateau solution $M_b := M(a,b,\varphi)$ for each $b>0$. We claim the function $p(b) := p(M_b)$ is continuous in b. To see this, we take two converging sequences (b_l) and (b_k) with the same limit b_0. If $\lim p(b_l) \neq \lim p(b_k)$ this would contradict the uniqueness of the minimal section, since the corresponding minimal surfaces also converge: There is a uniform curvature estimate for each sequence of minimal surfaces M_{b_l} by [GK10]. Moreover since the surfaces represent sections, each sequence satisfies a uniform area estimate. Writing the surfaces as graphs in normal coordinates over their tangent planes, we see that the sequence contains a converging subsequence with limit $M = M_{b_0}$.

We will now show that there exists $b_t \in \mathbb{R}$ such that $p(b) < 0$, for all $b > b_t$, and $\lim_{b\to 0} p(b) > 0$. By the intermediate value theorem this proves the lemma.

- To define b_t, we consider the horizontal helicoid M_H constructed by Daniel and Hauswirth [DH09] from Section 10.2, whose horizontal axis coincides with c_1. By Lemma 10.1 there exists a helicoid M_H for each pitch $a>0$ such that the incident vertical arcs of Γ are contained in its rulings. Since $\varphi < \pi/2$ the helicoid M_H and M_b are both minimal sections over a bounded convex domain $\pi(M_H) \cap \pi(M_b)$, whose boundary consists of three geodesic arcs.

  We claim there exists $b_t > 0$, such that the surfaces intersect in Γ only. In Section 10.2 we showed that the conormal η_H of the helicoid along the vertical ruling is horizontal and its opening angle depends continuously on the height given by

$$h = \frac{\sinh\left(\frac{-1}{2}\ln\left(\frac{1+\cos\varphi}{1-\cos\varphi}\right)\right)}{2\alpha^2},$$

where α depends on the pitch a, see Lemma 10.1. We consider the vertical plane V, given by its tangent plane, that is spanned by the conormal η_H and ξ in height h. Since each conormal is horizontal and turns monotonically, the intersection $V \cap M_H$ meets the vertical ruling in exactly one point given by the height h. Moreover M_H is a section in the interior and therefore $V \cap M_H$ is bounded from below. Hence for $a > 0$ and $\varphi \in (0, \pi/2)$ exists $b_t \geqslant |h|$ such that the helicoid is a barrier lying above M_b for every $b > b_t$.

Consequently we can estimate the vertical component of the conormal η of the Plateau solution M_b by the helicoid conormal η_H along the interior of the curve c_1:

$$\langle \eta, \xi \rangle < \langle \eta_H, \xi \rangle \quad \text{and hence} \quad p(M_b) = \int \langle \eta, \xi \rangle < \int \langle \eta_H, \xi \rangle = 0.$$

- Otherwise for $b \to 0$ we consider a minimal surface N in $\mathrm{Nil}_3(\mathbb{R})$ defined in [GK10, Section 5.], see Section 10.3 with the parameters a and φ. It has a positive period p_N, since it is bounded from below by a horizontal umbrella.

  For $b \to 0$ we have $M_b \to N$ away from the singularity. Furthermore, the sequence of conormals η_b converges uniformly to η_N on compact sets $K \subset c_1$. Therefore, the period $p(b)|_K$ converges uniformly to $p_N|_K > 0$ for each compact $K \subset c_1$ and $b \to 0$. Hence, on c we have $p(b) > 0$ for $b \to 0$.

□

Remark 11. For $\varphi = \pi/2$ the proof does not work since the helicoid M_H is not a barrier for $b < \infty$. Therefore, we cannot construct a genus 1 catenoid in $\mathbb{H}^2 \times \mathbb{R}$ with this method.

Lemma 11.9. *For each $a > 0$ and $\varphi \in (0, \pi/2)$ the map $b\colon \mathbb{R}_+ \times (0, \pi/2) \to \mathbb{R}_+$ defined by Proposition 11.8 is a continuous function.*

Proof. We assume that b is discontinuous in (a, φ), i.e. there exist two sequences (a_l, φ_l) and $(\bar{a}_l, \bar{\varphi}_l)$ with limit (a_0, φ_0) but w.l.o.g. $b_0 = \lim b(a_l, \varphi_l) > \lim b(\bar{a}_l, \bar{\varphi}_l) = \bar{b}_0$. There is a uniform curvature estimate for each sequence of minimal surfaces $M(a_l, b(a_l, \varphi_l), \varphi_l)$ and $M(\bar{a}_l, b(\bar{a}_l, \bar{\varphi}_l), \bar{\varphi}_l)$ by [GK10]. Moreover since the surfaces represent sections, each sequence satisfies a uniform area estimate. Writing the surfaces as graphs in normal coordinates over their tangent planes, we see that each one contains a converging subsequence with limit $M = M(a_0, b_0, \varphi_0)$ and $\bar{M} = M(a_0, \bar{b}_0, \varphi_0)$ respectively. Both minimal surfaces M and $\bar{M}$ have zero period. But for $b_0 > \bar{b}_0$ the minimal surface M bounds a mean convex domain from above with $\partial\bar{M}$ in its boundary. Therefore, M is an upper barrier for $\bar{M}$ with $\langle \eta, \xi \rangle > \langle \bar{\eta}, \xi \rangle$, contradicting the fact that both minimal surfaces have zero period. □

We solved the first period problem in b depending on (a, φ). Namely, for each angle $\varphi \in (0, \pi/2)$ and horizontal geodesic of length a, there exists an $b > 0$ as length of the vertical geodesic, such that the two horizontal mirror curves in the sister surface lie in the same mirror plane.

The second period problem relies on the horizontal mirror curve $\tilde{c}_2$ of length b and there are two difficulties, we have to solve: First of all, we have to make sure that the two vertical mirror planes, which are perpendicular to $\tilde{c}_2$ intersect. And secondly their angle of intersection has to be π/k. For the answers we need to restate the solution of the first period problem: For an angle φ and a vertical geodesic c_2 of length b there exists a horizontal geodesic of length a, such that the period is zero:

Proposition 11.10. *For each $b > 0$ and $\varphi \in (0, \pi/2)$ there exists $a(b, \varphi) > 0$, such that the horizontal period of the minimal surface $M(a(b, \varphi), b, \varphi)$ is zero.*

Proof. By Lemma 11.9 the map $b\colon \mathbb{R}_+ \times (0, \pi/2) \to \mathbb{R}_+$ continuous. For each φ we claim $b(a, \varphi) \to 0$ for $a \to 0$. Indeed, by the proof of Proposition 11.8 there exists $-h \in \mathbb{R}$ as an upper bound of $b(a, \varphi)$ given by a and φ:

$$-h = \frac{\sinh\left(\frac{1}{2}\ln\left(\frac{1+\cos\varphi}{1-\cos\varphi}\right)\right)}{2\alpha(a)^2} \geqslant b(a, \varphi), \quad \text{since otherwise} \quad p > 0.$$

By Lemma 10.1 we know that $\alpha(a) \to \infty$ for $a \to 0$. Therefore, the height h converges to zero as well as $b(a, \varphi)$ converges to zero.

Remains to show that the map b is unbounded. For each $\varphi > 0$ assume the contrary: The function $b_\varphi(a) := b(a, \varphi) \leqslant \hat{b}$ is bounded. Since b is a continuous function, we have $p(M_{\tilde{b}}) \neq 0$ for $\tilde{b} > \hat{b}$. The minimal surface $M(a, b(a, \varphi), \varphi)$ is a barrier from above for $M(a, \tilde{b}, \varphi)$, therefore $p(M_{\tilde{b}}) < 0$. The continuity of p in b implies $p(M_{\tilde{b}}) < 0$ for $\tilde{b} > \hat{b}$ and for all $a > 0$.

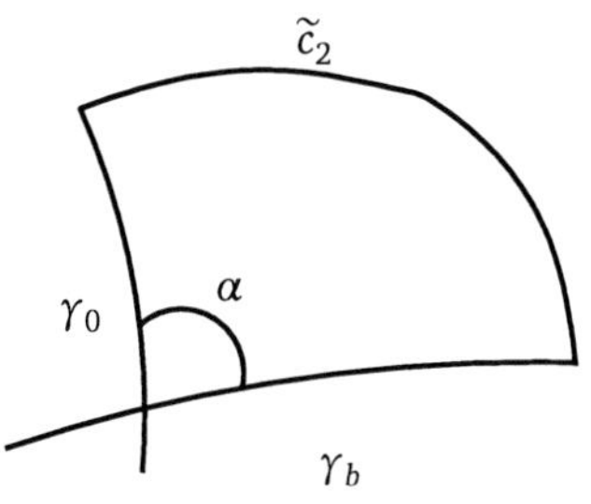

Figure 11.2.: Sketch of the defined geodesics γ_i, $i = 0, b$.

Consider the map $p_1 \colon a \mapsto p(M(a, \widetilde{b}, \varphi))$. We claim the function p_1 is continuous. To see this, we take two converging sequences (a_l) and (a_k) with the same limit a_0. Since the corresponding minimal surfaces also converge, this implies $\lim p(a_l) = \lim p(a_k)$. Moreover, for a large enough the minimal surface has a positive period,therefore there exists $\hat{a} > 0$ with $p_1(\hat{a}) = 0$, which is a contradiction.

Hence, $b_\varphi(a)$ is unbounded with $b_\varphi(a) \to 0$ for $a \to 0$. So we conclude: For all $b > 0$ there exists $a(b, \varphi) > 0$ (not necessarily unique) such that $M(a(b, \varphi), b, \varphi)$ has zero period. □

Remark 12. For conjugate minimal surfaces in $\mathbb{R}^3$ we have a unique $a(b, \varphi)$, since $a \mapsto b(a, \varphi)$ is injective because of scaling.

Before we solve the second period problem, we want to analyse the finite horizontal mirror curve $\widetilde{c}_2$ in $\mathbb{H}^2 \times \mathbb{R}$ parametrized by arc length. We choose the up-pointing surface normal ν and (c_2', η, ν) positive oriented. Where c_2 is the sister curve, then $\langle c_2', \xi \rangle = -1$. We consider the twist $\varphi(t)$ of c_2. The curvature of $\widetilde{c}_2$ is $\widetilde{k} = 1 - \varphi'(t) \leqslant 1$, since $\varphi' \geqslant 0$ by graph property of the minimal surface. Proposition 7.11 implies the embeddedness, since $\vartheta \leqslant \varphi < \pi/2$; moreover, $\vartheta \geqslant 0$ since $\vartheta < 0$ implies $\widetilde{k} > 1$. With γ_0 and γ_b we denote the unique geodesics given by $\gamma_i(0) = \widetilde{c}_2(i)$ and $\gamma_i'(0) = -\widetilde{\nu}(i)$, $i = 0, b$, see Figure 11.2.

As said before, in a first step we have to make sure, that the geodesics intersect:

Proposition 11.11. *For each $\varphi \in (0, \pi/2)$ exists $b_0 := b_0(\varphi) > 0$ satisfying*

$$b_0 e^{b_0} = \frac{1 - \cos(\varphi - b_0)}{\sin(\varphi - b_0)}, \quad b_0(\varphi) < \varphi$$

such that the vertical mirror planes of $\widetilde{M}(a(b,\varphi),b,\varphi)$ intersect for all $b \in (0, b_0(\varphi))$ and define an intersection-angle $\alpha > 0$.

Proof. As in Proposition 7.11 we consider the foliation by horocycles given by $\widetilde{v}(0)$ and the related angle $\vartheta \leqslant \varphi < \pi/2$. For the calculation we consider the upper half-plane and orient $\widetilde{c}_2$ such that $\widetilde{c}_2(0) = (0,1)$ and $\widetilde{v}(0) = (0,1)$, then γ_0 is contained in the y-axis. We want to parametrize the unique geodesic $\gamma_b \subset \mathbb{H}^2$, which starts in the endpoint of $\widetilde{c}_2$ ($\widetilde{c}_2(b) = (c_x, c_y)$) and its tangent is $(-\sin(\pi-\vartheta), \cos(\pi-\vartheta))$. For $\vartheta \neq k\pi, k \in \mathbb{Z}$, γ_b is an Euclidean halfcircle with radius r and its midpoint on the x-axis. We solve the linear equations

$$\gamma_b(\pi-\vartheta) = (c_x, c_y) = (x + r\cos(\pi-\vartheta), r\sin(\pi-\vartheta)).$$

The geodesic is in Euclidean coordinates parametrised by

$$\gamma_b(t) = \left(c_x + c_y\left(\frac{\cos\vartheta - \cos t}{\sin\vartheta}\right), c_y\frac{\sin t}{\sin\vartheta}\right).$$

The geodesics intersect if the x-coordinate of $\gamma_b(\pi)$ is negative:

$$c_x + c_y\frac{\cos\vartheta - 1}{\sin\vartheta} < 0. \tag{11.1}$$

Since $d_{\mathbb{H}^2}(\widetilde{c}_2(0), \widetilde{c}_2(b)) \leqslant b$ we have $c_x \leqslant b$ and $c_y > e^{-b}$. So we conclude Equation (11.1) is true if

$$\frac{1-\cos\vartheta}{\sin\vartheta} > be^b.$$

The angles ϑ and φ are related by $\vartheta' = \varphi' + \cos\vartheta - 1$ (see Proposition 7.11) and $\cos\vartheta \geqslant 0$ implies $\int(\cos\vartheta - 1) \geqslant -b$. Therefore we know that $\vartheta \geqslant \varphi - b$. Furthermore, the function $\vartheta \mapsto (1-\cos\vartheta)/\sin\vartheta$ increases monotonically, so we conclude, the geodesics intersect if

$$be^b < \frac{1-\cos(\varphi-b)}{\sin(\varphi-b)}.$$

This is equivalent to

$$f_\varphi(b) := \frac{1-\cos(\varphi-b)}{\sin(\varphi-b)} - be^b > 0. \tag{11.2}$$

Its differential

$$f'_\varphi(b) = \frac{\cos(\varphi - b) - 1}{\sin^2(\varphi - b)} - e^b(1 + b)$$

is less than zero for all $b < \varphi$. Hence, f_φ is decreasing. Let us consider the limits at the boundaries:

$$\lim_{b \to 0} f_\varphi(b) = \frac{1 - \cos\varphi}{\sin\varphi} > 0, \qquad \text{for } \varphi \in (0, \pi/2)$$
$$\lim_{b \to \varphi} f_\varphi(b) = -\varphi e^\varphi < 0.$$

We conclude, there exists exactly one $b_0(\varphi) \in (0, \varphi)$ with $f_\varphi(b_0(\varphi)) = 0$. Moreover, for all $b < b_0(\varphi)$ we have $f_\varphi(b) > 0$, therefore the geodesics intersect for all $b < b_0(\varphi)$. □

Remark 13. For $b = \varphi$ the total curvature of $\tilde{c}_2$ is $b - \varphi = 0$ and therefore the two geodesics γ_0 and γ_b do not intersect in any $p \in \mathbb{H}^2$, but in $\partial\mathbb{H}^2$. By Proposition 11.10 there exists $a > 0$ to solve the first period problem, hence after reflection the contruction causes a complete singly periodic MC-1/2 surface $\widetilde{M}(a(\varphi,\varphi),\varphi,\varphi)$ in $\mathbb{H}^2 \times \mathbb{R}$ with infintely many ends.

Now we are able to solve the second period problem that is given by the angle α. We want the surface to close after $2k$ reflections about vertical mirror planes, so α has to be π/k.

Proposition 11.12. *For each $k \geqslant 3$ there exists $\epsilon = \epsilon(k) > 0$, such that $\varphi = \pi/k + \epsilon < \pi/2$ and there exists $0 < b < b_0(\varphi)$, such that the surface $\widetilde{M}(a(b,\varphi), b, \varphi)$ has angular period $\alpha = \pi/k$ and $p(\widetilde{M}(a(b,\varphi), b, \varphi)) = 0$.*

Proof. For $b < b_0(\varphi)$ the geodesics γ_0 and γ_b intersect by Proposition 11.11. Hence, we can apply the Gauß-Bonnet Theorem to the compact disc $V \subset \mathbb{H}^2$ defined by $\partial V = \tilde{c}_2 \cup \gamma_0 \cup \gamma_b$:

$$\int_V K + \int_{\partial V} k_g + \sum \alpha_i = 2\pi\chi(V),$$

where α_i, $i = 1, 2, 3$ are the exterior angles with $\alpha_i = \pi/2$ for $i = 1, 2$ and $\alpha_3 = \pi - \alpha$. Furthermore, k_g is the geodesic curvature with respect to the inner normal of ∂V and since $\tilde{c}_2$ is a mirror curve $k_g = -\tilde{k}$ with respect to the surface normal. Using this we get

$$\int_V K - \int_{\tilde{c}_2} \tilde{k} + 2\pi - \alpha = 2\pi. \tag{11.3}$$

From Lemma 7.9 we know $\tilde{k} = 1 - \varphi'$. Integrating this shows Equation (11.3) is equivalent to $\varphi - b - \mathrm{area}(V) = \alpha$.

We claim, that the lengths $l(\gamma_i)$ are bounded from above for all $b \leqslant b_0(\varphi)$. Recall from the proof of Proposition 11.11 that $\varphi \geqslant \vartheta \geqslant \varphi - b$. In particular, we have a lower bound for ϑ, the angle that is given by the tangent of $\tilde{c}_2$ and the horocycle fibration. Assume that the lengths $l(\gamma_i)(b) \to \infty$ for $b \to 0$, this implies $\vartheta \to 0$, a contradiction. If the lengths are bounded, the area tends to zero for $b \to 0$.

The angle α depends continuously on $b < b_0(\varphi)$:

$$\alpha(b) = \varphi - b - \mathrm{area}(V(b)) \tag{11.4}$$

and decreases. The idea is to show that for any $k \geqslant 3$ exists $\varphi_k \in (0, \pi/2)$ such that

$$\lim_{b\to 0} \alpha(b) > \pi/k \quad \text{and} \quad \lim_{b\to b_0(\varphi_k)} \alpha(b) < \pi/k.$$

On the one hand $\lim_{b\to 0} \alpha(b) = \varphi_k$, hence we have to choose $\varphi_k > \pi/k$. Therefore, for any $\epsilon_k > 0$, $\varphi_k := \pi/k + \epsilon_k$ satisfies the first condition.

On the other hand we have

$$\lim_{b\to b_0(\varphi_k)} \alpha(b) = \varphi_k - b_0(\varphi_k) - \mathrm{area}(V(b_0(\varphi_k))),$$

hence we have to choose $\epsilon_k > 0$ such that

$$\varphi_k < \frac{\pi}{k} + b_0(\varphi_k) + \mathrm{area}(b_0(\varphi_k)) \quad \Leftrightarrow \quad \epsilon_k < b_0(\varphi_k) + \mathrm{area}(b_0(\varphi_k)).$$

We claim the function $b_0(\varphi)$ increases monotonically. Recall Equation (11.2): b_0 was defined implicitly by $f_\varphi(b) = \frac{1-\cos(\varphi-b)}{\sin(\varphi-b)} - be^b = 0$. By chain rule we have

$$b_0'(\varphi) = e^{b_0(\varphi)}(1 + b_0(\varphi)) + \frac{1-\cos(\varphi - b_0(\varphi))}{\sin^2(\varphi - b_0(\varphi))} > 0.$$

Therefore with $\epsilon_k = b_0(\pi/k)$ we get:

$$\lim_{b\to b_0(\varphi_k)} \alpha(b) = \pi/k + \underbrace{b_0(\pi/k) - b_0(\varphi_k)}_{<0} - \mathrm{area}(V(b_0(\varphi_k))) < \pi/k.$$

Remains to show that $\varphi_k < \pi/2$ for all $k \geqslant 3$. Since b_0 increases this is true if $\pi/3 + b_0(\pi/3) < \pi/2$, i.e. $b_0(\pi/3) < \pi/6$. But this follows directly from the fact that $f_{\pi/3}(\pi/6) < 0$.

By the intermediate value theorem follows, there exists $b^* \in (0, b_0(\pi/k + \epsilon_k))$ such that $\alpha(b^*) = \pi/k$. By Proposition 11.10 there exists $a > 0$ to solve the first period problem. □

The proposition proves the existence of one surface $M(a_k, b_k, \varphi_k)$ for any $k \geqslant 3$. It is natural to ask if there is a family of CMC surfaces with the desired symmetries and k ends. The answer is yes:

Proposition 11.13. *For each $k \geqslant 3$ there exists an interval $U_k \subset (0, \pi/2)$, such that for all $\varphi \in U_k$ there exists $b(\varphi) > 0$, such that the surface $\widetilde{M}(a(b(\varphi), \varphi), b(\varphi), \varphi)$ has angular period $\alpha = \pi/k$ and $p(\widetilde{M}(a(b, \varphi), b, \varphi)) = 0$.*

Proof. In Proposition 11.12 the existence of a triple $(a(b(\varphi_k), \varphi_k), b(\varphi_k), \varphi_k)$ was proven such that the surface $\widetilde{M}(a(b(\varphi_k), \varphi_k), b(\varphi_k), \varphi_k)$ has the desired properties. By Equation (11.4) we know, the pair $(b(\varphi_k), \varphi_k)$ is a zero of the following continuously differentiable function

$$G(b, \varphi) = \alpha(b, \varphi) - \pi/k = \varphi - b - \operatorname{area}(V(b, \varphi)) - \pi/k.$$

The angle $\alpha(b, \varphi)$ is given by the two geodesics γ_0 and γ_b. Recall from the proof of Proposition 11.12 that $\partial_b \alpha(b, \varphi) < 0$. By the implicit value theorem there exist an open neighborhood U_0 of φ_k, an open neighborhood V of $b(\varphi_k)$, and a unique continuously differentiable function $g \colon U_0 \to V$ with $g(\varphi_k) = b(\varphi_k)$ such that $G(g(\varphi), \varphi) = 0$ for all $\varphi \in U_0$.

To define U_k notice that $g(\varphi_k) < b_0(\varphi_k)$, therefore the subset $U_k := \{\varphi \in U_0 \colon g(\varphi) < b_0(\varphi)\} \cap (0, \pi/2)$ is not empty. Hence, for all $\varphi \in U_k$ there exists $b = g(\varphi) < b_0(\varphi)$ and therefore by Proposition 11.10 there exists $a > 0$ to solve the first period problem. □

Remark 14. We want to analyse the limiting cases:

- $\varphi \to \inf U_k$: We know that $\inf U_k \geqslant \pi/k$. From $\varphi \to \pi/k$ follows $b + \operatorname{area}(V(b, \varphi)) \to 0$, which implies $b \to 0$. Assuming the solution of the first period problem $a(b, \varphi) > 0$ for $b \to 0$ leads to the k-noid from [GK10] which has a positive period. Therefore if $\inf U_k = \pi/k$ then the sequence of k-noids converges to an union of k horocylinders away from the singularity for $\varphi \to \pi/k$.

- $\varphi \to \sup U_k \leqslant \pi/2$: Since $\partial_\varphi \operatorname{area}(V(b, \varphi)) \leqslant 0$, $\varphi \to \sup U_k$ implies that b increases. For the CMC surfaces this means that the length of the finite horizontal symmetry curve grows.

11.2.4 Main Theorem

After solving the two period problems we can now prove the existence of the MC 1/2 surface with genus 1:

Theorem 11.14. *For $k \geqslant 3$, there exists a family of surfaces M with constant mean curvature $1/2$ in $\mathbb{H}^2 \times \mathbb{R}$ such that:*
- *M is a proper immersion of a torus minus k points,*
- *M is Alexandrov embedded.*
- *M has k vertical mirror planes enclosing an π/k-angle,*
- *M has one horizontal mirror plane.*

Proof. By [Dan07] the fundamental piece $M(a, b_0)$ has a sister surface $\widetilde{M}$ with constant mean curvature $1/2$ in $\mathbb{H}^2 \times \mathbb{R}$, which is a graph. By construction and from the solution of the period problems, $\widetilde{M}$ has one horizontal and two vertical mirror planes; the two vertical mirror planes enclose an angle π/k. It consists of 4 mirror curves: two horizontal (one bounded and one unbounded) and two vertical (one bounded and one unbounded). After Schwarz reflection about the vertical infinite mirror curve followed by reflection about the horizontal mirror plane we have the fundamental domain of one end: It is built up of four fundamental domains $\widetilde{M}$. We use the Euler characteristic

$$\chi = V - E + F = 2 - 2g$$

to determine the genus g of the complete MC 1/2 surface $\widetilde{M}$ with k ends, which is generated by Schwarz reflection. We have $\chi = 4k - 8k + 4k = 0$ and therefore $g = 1$.

The Alexandrov-embeddedness follows from analogous arguments as in the $2k$-noid case. □

Conclusion and outlook

Let us summarise the main results of this thesis.

We constructed new constant mean curvature surfaces in homogeneous 3-manifolds. They arise as sister surfaces of Plateau solutions.

The first example, a two-parameter family of MC H surfaces in $\Sigma(\kappa) \times \mathbb{R}$ with $H \in [0, 1/2]$ and $\kappa + 4H^2 \leqslant 0$, has genus 0, $2k$ ends and k-fold dihedral symmetry, $k \geqslant 2$. The existence of the minimal sister follows from the construction of a mean convex domain. The projection of the domain is non-convex.

The second example is an MC $1/2$ surface in $\mathbb{H}^2 \times \mathbb{R}$ with k ends, genus 1 and k-fold dihedral symmetry, $k \geqslant 3$. We had to solve two period problems in the construction. The first period guarantees that the surface has exactly one horizontal symmetry. For the second period we had to control a horizontal mirror curve to get the dihedral symmetry. In the case of $H \neq 0$ the total curvature of a horizontal mirror curve depends not only on the twist of the normal, but also on its length.

For $H = 1/2$ all the surfaces are Alexandrov-embedded. An interesting problem is to construct non-embedded examples that correspond to those presented here. Große-Brauckmann proved in [Gro93] that each Delaunay surface is the associated MC 1 surface of a helicoid in $\mathbb{S}^3$. The family of Delaunay surfaces consists of embedded (undoloid) and non-embedded (nodoid) examples. It would be natural to construct nodoid type examples of k-noids in $\mathbb{H}^2 \times \mathbb{R}$. In the case of a nodoid the vertical component of the normal $\langle \widetilde{\xi}, \widetilde{\nu} \rangle$ changes the sign. Hence, the sister surface is not a graph anymore. Beside this it would be interesting to study the geometry of the ends for $H < 1/2$ in general.

Recall that in the construction of the surface with genus 1, we had to constrain the angle in the projection, $\varphi < \pi/2$. In $\mathbb{R}^3$ Schoen proved that the unique properly embedded minimal surface of finite topology with two ends is the catenoid, hence has genus zero. See [Sch83]. The construction in Section 11.2 suggests that the number of ends is sharp.

Bibliography

[Alm03] Sergio de Moura Almaraz, *Geometrias de Thurston e Fibrados de Seifert*, Master's thesis, Pontifícia Universidade Católica do Rio de Janeiro, 2003.

[AR05] Uwe Abresch and Harold Rosenberg, *Generalized Hopf differentials*, Mat. Contemp. **28** (2005), 1–28 (English).

[CM11] Tobias H. Colding and William P. Minicozzi II, *A course in minimal surfaces*, Providence, RI: American Mathematical Society (AMS), 2011 (English).

[Dan07] Benoît Daniel, *Isometric immersions into 3-dimensional homogeneous manifolds*, Comment. Math. Helv. **82** (2007), no. 1, 87–131 (English).

[DH09] Benoît Daniel and Laurent Hauswirth, *Half-space theorem, embedded minimal annuli and minimal graphs in the Heisenberg group*, Proceedings of the London Mathematical Society **98** (2009), no. 2, 445–470.

[Dou31] Jesse Douglas, *Solution of the problem of Plateau*, Trans. Am. Math. Soc. **33** (1931), 263–321 (English).

[Esc89] Jost-Hinrich Eschenburg, *Maximum principle for hypersurfaces*, Manuscr. Math. **64** (1989), no. 1, 55–75 (English).

[GHL04] Sylvestre Gallot, Dominique Hulin, and Jacques Lafontaine, *Riemannian geometry*, 3rd ed., Universitext, Springer, 2004 (English).

[GK09] Karsten Große-Brauckmann and Robert B. Kusner, *Ruled minimal surfaces in homogeneous manifolds and their sisters*, preprint, 2009.

[GK10] ———, *Conjugate Plateau constructions for homogeneous 3-manifolds*, preprint, 2010.

[GKS03] Karsten Große-Brauckmann, Robert B. Kusner, and John M. Sullivan, *Triunduloids: embedded constant mean curvature surfaces with three ends and genus zero*, J. Reine Angew. Math. **564** (2003), 35–61 (English).

[GKS07] ———, *Coplanar constant mean curvature surfaces*, Commun. Anal. Geom. **15** (2007), no. 5, 985–1023 (English).

[Gro93] Karsten Große-Brauckmann, *New surfaces of constant mean curvature*, Math. Z. **214** (1993), no. 4, 527–565 (English).

[GT01] David Gilbarg and Neil S. Trudinger, *Elliptic partial differential equations of second order*, reprint of the 1998 ed., Berlin: Springer, 2001 (English).

[Gul73] Robert D. Gulliver, *Regularity of minimizing surfaces of prescribed mean curvature*, The Annals of Mathematics **97** (1973), no. 2, pp. 275–305 (English).

[HRS08] Laurent Hauswirth, Harold Rosenberg, and Joel Spruck, *On complete mean curvature $1/2$ surfaces in $\mathbb{H}^2 \times \mathbb{R}$*, Commun. Anal. Geom. **16** (2008), no. 5, 989–1005 (English).

[HS88] Joel Hass and Peter Scott, *The existence of least area surfaces in 3-manifolds*, Trans. Am. Math. Soc. **310** (1988), no. 1, 87–114 (English).

[KN63] Shoshichi Kobayashi and Katsumi Nomizu, *Foundations of differential geometry. I*, New York-London: Interscience Publishers, a division of John Wiley & Sons Inc., 1963 (English).

[Law70] H. Blaine Jr. Lawson, *Complete minimal surfaces in $\mathbb{S}^3$*, The Annals of Mathematics **92** (1970), no. 3, pp. 335–374 (English).

[Mil76] John W. Milnor, *Curvatures of left invariant metrics on Lie groups*, Adv. Math. **21** (1976), 293–329 (English).

[Mor66] Charles Bradfield jun. Morrey, *Multiple integrals in the calculus of variations*, Die Grundlehren der mathematischen Wissenschaften, no. 130, Berlin-Heidelberg-New York: Springer-Verlag, 1966 (English).

[MP11] William H. Meeks III and Joaquín Pérez, *Constant mean curvature surfaces in metric Lie groups*, Contemporary Mathematics (to appear), 2011.

[MT11] José Miguel Manzano and Francisco Torralbo, *New examples of constant mean curvature surfaces in $\mathbb{S}^2 \times \mathbb{R}$ and $\mathbb{H}^2 \times \mathbb{R}$*, arXiv:1104.1259v2 [math.DG], 2011.

[MY82] William H. Meeks III and Shing-Tung Yau, *The existence of embedded minimal surfaces and the problem of uniqueness*, Mathematische Zeitschrift **179** (1982), 151–168, 10.1007/BF01214308.

[Nit75] Johannes C.C. Nitsche, *Vorlesungen über Minimalflächen*, Die Grundlehren der mathematischen Wissenschaften, no. 199, Berlin-Heidelberg-New York: Springer, 1975 (German).

[NR06] Barbara Nelli and Harold Rosenberg, *Simply connected constant mean curvature surfaces in* $\mathbb{H}^2 \times \mathbb{R}$, Mich. Math. J. **54** (2006), no. 3, 537–543 (English).

[Oss70] Robert Osserman, *A proof of the regularity everywhere of the classical solution to Plateau's problem*, The Annals of Mathematics **91** (1970), no. 3, pp. 550–569 (English).

[Rad30] Tibor Radó, *The problem of the least area and the problem of Plateau*, Mathematische Zeitschrift **32** (1930), 763–795 (English).

[Sch83] Richard Schoen, *Uniqueness, symmetry, and embeddedness of minimal surfaces.*, J. Differ. Geom. **18** (1983), 791–809 (English).

[Sco83] Peter Scott, *The geometries of 3-manifolds*, Bull. Lond. Math. Soc. **15** (1983), 401–487 (English).

[Thu97] William P. Thurston, *Three-dimensional geometry and topology*, vol. 1, Princeton Mathematical Series, no. 35, Princeton, NJ: Princeton University Press, 1997 (English).

Index

Curriculum vitae

Personal information

Name	Julia Plehnert
Date of birth	13 November 1981
Place of birth	Bremen

Employment

since 2008	Research assistant, Technische Universität Darmstadt
2006 – 2008	Teaching assistant, Technische Universität Darmstadt
2003 – 2005	Student assistant, Fraunhofer Institut, IPSI, Darmstadt

Academic studies

2008 – 2012	Doctoral Studies, Technische Universität Darmstadt Mathematics
2006 – 2008	Master of Science, Technische Universität Darmstadt Mathematics
2005	Exchange student, Unicamp, Campinas, Brazil
2002 – 2006	Bachelor of Science, Technische Universität Darmstadt Mathematics with computer science

Education

1994 – 2001	Max-Planck-Gymnasium Delmenhorst Abitur
1992 – 1994	Wilhelm-von-der-Heyde-Schulzentrum, Delmenhorst Orientierungsstufe
1989 – 1992	Parkschule Delmenhorst Grundschule